Born in the UK in 1980, Angela Saini is a science journalist and reports for the BBC, *New Scientist*, *Wired* and the *Economist*. She was named European Young Science Writer of the Year in 2009, and in 2008 won a European television news award.

GEEK NATION

HOW INDIAN SCIENCE IS
TAKING OVER THE WORLD

ANGELA SAINI

HODDER

First published in Great Britain in 2011 by Hodder & Stoughton
An Hachette UK company

First published in paperback in 2012

1

Copyright © Angela Saini 2011

A CIP catalogue record for this title is available from the British Library

ISBN 978 1 444 71016 8

Illustrations by Hello Marine

Typeset by Palimpsest Book Production Limited,
Falkirk, Stirlingshire

Printed and bound by Clays Ltd, St Ives plc

Hodder & Stoughton policy is to use papers that are natural,
renewable and recyclable products and made from wood grown in
sustainable forests. The logging and manufacturing processes are expected
to conform to the environmental regulations of the country of origin.

Hodder & Stoughton Ltd
338 Euston Road
London NW1 3BH

www.hodder.co.uk

For my sisters, Monica and Rima

CONTENTS

PREFACE

In 1881 in the village of Bakhshali, in a northern province of British-ruled India, a farmer was digging around some stone ruins in a field when he made a discovery.

It was a few tattered sheets of birch bark, each inscribed with tiny black writing in a dialect so old that it hadn't been seen around here, let alone used, for a thousand years. But there was something about them that was even more significant than their age. Unlike other medieval texts recovered from Europe and China, this wasn't a work of religion, philosophy or art. The pages had unusual symbols on them, some written vertically, some horizontally, and all separated by thick lines.

It was a mathematics textbook.

It would later become known as the Bakhshali Manuscript. Historians believe that it may have been written around AD 700, making it one of the earliest surviving scientific texts in Asia. At that time Indians were using decimal points, square roots and algebra, centuries before the West had heard of them. Starting two hundred years earlier, the mathematicians Aryabhata, Bhaskara and Brahmagupta developed the numbers we use today. In fact Brahmagupta was the first to popularise the zero, allowing equations to be written that would come to underpin all of mathematics. And Indian astronomers are also thought to have been the first to calculate that the earth spins on its own axis to create day and night.

But of course in 1881, Bakhshali was just another farming village in an impoverished country with not much more to offer the world than tea, cotton and spices. The European Enlightenment had happened. Modern science, using experiments to test hypotheses, hadn't emerged in India but in the Royal Society in London. The zero was being used in the West, its origins ignored. And like the appropriated numbers, the Bakhshali Manuscript also left India, bound for England.

It ended up in the University of Oxford. It lies there to this day, stored behind glass in the famous Bodleian Library. In the century since it was dug up by the farmer, the United States, Europe and Japan have all become global superpowers, while India has been left behind. Only now have people started asking whether this might change and Asian countries might become powerful economic giants once more, reclaiming the scientific and technological legacy they lost. For those in high-tech hubs like Silicon Valley who doubt that this is possible, crumbling away on the Bodleian's dark, wooden shelves, surrounded by dusty tomes and portraits of British kings and queens, this ancient textbook is a silent reminder that India was once one of the most powerful scientific nations of all, that it may even be the birthplace of the geek.

INTRODUCTION

They say that history is cyclical.

That's the thought in my mind when I travel from England to south India, catching a ride at the end on a bus going to the Vikram Sarabhai Space Centre, the government's biggest scientific facility researching the stratosphere and beyond. It's in the fishing village of Thumba, at the end of a bumpy road lined with candy-coloured shacks and surrounded by coconut groves.

A moustachioed minder eyes my bag suspiciously and reminds me what to expect. 'It's top secret, OK?' he says. 'Have you got your pass? No cameras allowed. And no other electronics, too.' This is the fourth time he's told me.

I reassure him, again, that I understand, clinging onto the sweat-stained plastic seat cover with my fingernails as we bounce into another pothole. As the bus rumbles along, I see a tailor on the left, working with a black iron sewing-machine in an otherwise empty blue room. And on the right, I spot a small building, improbably named the 'National Typewriting and Computer Center'.

From the outside, the Space Centre looks like a prison. At the front it's surrounded by metal fencing and lookouts, while at the back is the Indian Ocean. The armed security guards, from the Central Industrial Security Force, speak dozens of Indian languages, from Urdu to Malayalam, so if you don't understand the orders they're barking at you, they just switch tongues until you do. Just as I had been warned, they empty my bag of its mobile phone, memory stick, voice recorder and iPod. They repeat this with a busload of excited scientists behind me, who have come here in a tour group from laboratories across India. It takes weeks to get permission but space officials occasionally allow visitors inside the centre for educational purposes, to give the public a peek at India's scientific achievements. I'm singled out for a brief interrogation: who am I, and what am I doing here?

I'm here to learn about this rediscovered nation of geeks. My dad, who worked as a chemical engineer in India in the 1960s, used to tell me about the great potential of this land of hardworking scientists and engineers. Yet India never managed to live up to his dreams – until now. The success of the Vikram Sarabhai Space Centre is among the first signs that India may have what it takes to become a scientific superpower in the same league as the United States, Europe and Japan. It seems like India is being pulled out of poverty and transformed into a technological giant.

But right now, like the scientists behind me, I'm just silently wishing that I might see a rocket launch.

We tiptoe into the vast facility like children at the gates of Disneyland. Ahead of us is nothing but rows of palm trees and a long sandy beach. Then without warning, a thundering 'whooooosh' blasts out over the sky, followed by a thin, wiggly contrail that zips over the ocean. We abandon our dignity and run towards the coast, craning our necks to spot it, but it's already gone. A visiting scientist from the central Indian state of Madhya Pradesh flicks her head around and puts her sunglasses back on. 'I've seen bigger crackers at Diwali,' she says wryly.

'It's only a climate rocket,' an amused guard tells me in Hindi. 'They use it to check the weather. It's launched from over there,' he adds, pointing a hundred metres to my left, behind a short stone building. The rocket was only slightly taller than me, around 30 centimetres in diameter and travelling 75 kilometres into the upper atmosphere.

There are too many people living nearby to safely launch big rockets from the beaches around Thumba, we're told. The largest ones are taken to a remote island around 700 kilometres to the northeast, next to the Bay of Bengal. That was the launch site for India's famous first mission to the moon in 2008. The small orbiter and probe, named *Chandrayaan-1* after the Sanskrit words meaning 'moon' and 'traveller', studied the lunar surface for months. Nobody was sure whether the project would be a success or a waste of time, but the reputation of Indian science depended on it. Many in the scientific community had always assumed that India would never be able to afford a space mission.

In the end, their ambitious project was completed at a cost of just 90 million dollars, five times less than the United States' most recent lunar orbiter. At first, people thought it had failed. It should have been on the moon for two years, but was forced to come back to earth after just ten months when it suddenly stopped sending back the radio signals that engineers on the ground needed to communicate with it.

Fortunately, its instruments had already done their job. Picking apart the data, researchers found scraps of evidence that there might be water on the moon, ending decades of speculation. According to the readings, there was ice in craters on the lunar north pole, dampish soils on its rocky surface and signs that still more water was being created. It brought closer the possibility that humans might eventually live on the moon, using the water for survival and to source the hydrogen fuel that might power spacecraft to travel to other planets. For the last few months, the scientists in Thumba have been celebrating.

And now the space programme has entered a new phase. According to newspaper reports, in around 2015 an Indian astronaut will be going into space for the first time in a rocket designed by Indian scientists.

Here on the ground, surrounded by palm trees and dirt roads, it must have seemed impossible a few decades ago, but then, things are changing fast. Not far away from the Space Centre, Indian chemists are producing lifesaving medicines that are being sold in Europe and Indian nerds are fixing computer software for Americans. This impoverished tea- and cotton-growing backwater is starting to reclaim the scientific legacy that it lost thousands of years ago. Staring into the clouds at the rocket, which has by now disappeared completely, I ask myself, how on earth did they do it?

The first clue lies in St Mary Magdalene Church, tucked away in the grounds of the Vikram Sarabhai Space Centre, facing the ocean. Around a century ago local Christian fishermen, from a line dating back to when India received its first foreign missionaries, found a sandalwood statue of Mary Magdalene washed up on the shore and gave the church its name. It's not a church any more, however. The only reminder of its religious history is a flowery inscription on a small plaque to the left of the entrance. 'In 1962,' it reads verbatim, 'the church authorities and the parishiners decided in a gracious and exemplary manner to dedicate this place of worship on the altar of science.'

In the 1960s, when engineers were building India's first space rocket, their budgets were so tight that they used this place as a makeshift laboratory, keeping spare parts elsewhere in tin-roofed sheds and launching small test rockets next to the empty sands nearby. Now, the church has been converted into a memorial to India's history in space. The altar is preserved inside, painted white with an orange stained-glass window in the shape of the sun. Everything else has been replaced with posters and model satellites. Rockets are propped up in the gardens out front like giant white tombstones. The biggest, a Polar Satellite Launch Vehicle, is the length of three buses.

Science has become a faith in its own right these days. I'm asked to take off my shoes before I enter the museum, as if it were a temple. In one corner, says the guide, I can see a replica of India's first satellite, launched in 1975. It's only a few metres across, octagonal and coated in blue solar panels. On the walls are black and white photographs of the people – prime ministers, presidents, engineers and physicists among them – who played a part in the early years of the space programme, each holding the secrets of India's meteoric rise.

But almost all of them died a long time ago, taking their stories with them. After weeks of phone calls, I finally track down one of the few people who worked at the space centre back then and who is still alive. Professor Udupi Ramachandra Rao is seventy-eight years old, making him one of the oldest rocket scientists in the world. But when I meet him, his face is as wrinkle-free as a toddler's and he still has a meticulous memory for details.

In a faint voice he tells me how his career began. It's the tale of how India, a country wracked for centuries by famine, poverty and illiteracy, managed to send a rocket to the moon.

'Arthur C Clarke, he was a great friend of mine, you know,' says Professor Rao, catching me scanning the titles on his bookshelf. His office is like a trophy cabinet, lined with wall-to-wall glass cupboards stuffed with honours and awards. Some are shaped like rockets and satellites; another is a plate with his face printed on it. On the shelf behind him, alongside a three-volume collection of his own writings, are books by Clarke, the late futurist, science-fiction writer and author of the famous novel-turned-film *2001: A Space Odyssey*. Rao is past retirement age but he still comes to this office at the headquarters of the Indian Space Research Organisation in a city to the north of the Vikram Sarabhai Space Centre, writing letters and cataloguing his memories.

He was a teenager in 1947 when India won independence from Great Britain, becoming the world's largest democracy. This was an age of possibilities and Rao's eyes were set on the stars. So he started by studying the particles that hit the earth's atmosphere from outer space, known as cosmic rays. At the time nobody had physically conquered the cosmos, but fortunately for Rao's career, the United States and the Soviet Union were about to begin the world's first space race. A Soviet satellite blasted off in October 1957, closely chased by a cosmonaut in 1961. The Americans retaliated by planning the first manned mission to the moon. Rao realised that if the future was indeed in space, that future was going to be either Russia's or America's. So in 1961 he packed his bags and found himself a job as an assistant professor at the University of Texas in Dallas.

While he was there, he got a call from the National Aeronautics and Space Administration, NASA, then only five years old. It was rare for foreign scientists to be allowed access to the secretive world of American space research. But the administration gave him funding to figure out how to make complicated space experiments work by testing them out using high-altitude balloons.

'I was in room number 26-441,' he recalls. He shared this office with an Australian scientist who would end up leaving to head up a space programme in his home country. 'And room 443 had two other characters,' he adds. By a twist of fate, this single corridor contained what was probably the greatest concentration of global space-science talent in history. The two researchers in room 443 each went on to lead the European and Japanese space agencies.

A similar future lay in store for Rao. He decided to head back to India on a promise from the government that he could help start the country's first space programme. It was a dream come true, he tells me. In 1969, less than a month after Neil Armstrong stepped on the moon, the Indian government founded the Indian Space Research Organisation, known as ISRO. And Rao worked there for the rest of his life, becoming chairman of the organisation in 1984.

In the foyer earlier, I'd passed a gleaming white replica of one of India's space rockets, the Geosynchronous Satellite Launch Vehicle. Rao's desk has its own mini rocket sitting next to a globe.

'I gave a talk once called Space, Next 1000 Years,' he says. He has a perfect memory but it comes with a habit of wandering off on tangents. 'Now, I said, we'll colonise Mars. And that is a difficult problem, but on other hand I think it will be done. Maybe another 500 years. Then, Arthur C Clarke may have said less. He was a visionary. For example, years ago he asked me, Rao, when do you think we'll see a space elevator? And I said minimum fifteen years, and he said, no, seven.'

'Really, seven?' I ask incredulously. Engineers have explored the idea of building a giant space elevator – a wire stretching from the earth to satellites in space – but they haven't come close to finding a material that can be made in large quantities that is also light and strong enough to withstand the stresses needed to stretch that far.

'I know! You're very crazy, I told him. But he was an extremely interesting person. Maybe we'll do it,' the professor says with a smile.

Rao obsesses about the future in a way that usually only science-fiction writers do. In the decades after he joined ISRO, both his experience at NASA and his friendship with Clarke helped him push the Indian space programme beyond its original limits. When critics said that India could never launch satellites or rockets, he was the one to prove them wrong. Indian leaders didn't believe they could send an orbiter to the moon, before he and his colleagues showed them they could. His heart still lies in that scientific golden age, when the apparently impossible came true; when man walked on the moon and scientists built a space station with shuttles taking astronauts back and forth.

Now Rao has his sights set on a new scientific age, in which India will be breaking barriers in research and developing the world's most futuristic technologies. Next to the textbooks on his bookshelf are copies of *The Complete Book of UFOs* and *Physics of Immortality*. Rao tells me that within thirty years, after Earth's raw materials start to run out, India will start mining resources on other planets.

Back in the 1960s, even building an Indian space rocket seemed impossible, let alone this fantastical prediction. Huge swathes of the country were living in poverty and two thirds of adults were illiterate. Both Indian politicians and the foreign countries on which India depended for food aid complained that space travel was an unnecessary extravagance.

Two men changed their minds. The first was Rao's boss, Vikram Sarabhai, the founder of India's space programme and the man after whom the space centre in Thumba is named. Born in 1919, Sarabhai had studied physics at Cambridge University and later married a famous Indian classical dancer. He was a good-looking geek with an electric personality.

Sarabhai urged Rao to come back to India from the US and work on the country's fledgling space programme. Even though he had never built a rocket, Sarabhai was such a towering force that the professor felt he couldn't refuse. 'When I came back to India, I didn't know much, but I knew I had to work under Sarabhai. Even though I was paid a very paltry sum. Everybody loved him; everyone felt they were near to him. He had this special quality,' says Rao.

But in the bigger scheme of things, Sarabhai was still just a physicist. And he needed cash. It was an old problem: Galileo Galilei would probably never have done his game-changing work in mathematics and astronomy without the help of the powerful Medici family of Rome, and an atom bomb would never have been built as quickly as it was if the US President Franklin Roosevelt hadn't pumped millions of dollars into the Manhattan Project. The Indian space programme needed this kind of political support. And it came in the shape of India's other important geek, Jawaharlal Nehru, India's first prime minister after the country gained independence from the British in 1947.

Like Sarabhai, Nehru was a charismatic Cambridge University science graduate. The two became friends. 'Sarabhai's family was politically connected. Not he himself involved in politics, but his sister was involved and his parents were involved. It was a very illustrious family,' explains Rao. And like Sarabhai, Nehru was convinced that science was the key to turning India's fortunes around.

One of the first things he did after becoming prime minister was to draw up a science policy and make himself the president of the Council for Scientific and Industrial Research, India's most productive research and development body. Invasion, colonisation, famine and partition had all but stripped the nation of its scientific legacy. There were only a few individual exceptions, such as the world-famous mathematician Srinivasa Ramanujan, and the physicists Jagadish Chandra Bose and Satyendra Nath Bose. But it had been Europeans, not Indians, who took advantage of centuries of progress in mathematics and engineering to build the first steam engines, aircraft, rockets and computers. As far as Nehru was concerned, it was India's greatest tragedy to have been wiped off the world map; building up its scientific base again was one way to get it back.

His dream was to have Indians not just working more in science, but also thinking more rationally in their everyday lives. India was a deeply religious country with high illiteracy rates, and swathes of the population were as superstitious as they had been before the Middle Ages. It wasn't unusual for people to show more faith in healing gurus than in doctors and dentists, or put their life decisions in the hands of an astrologer. This had to change, Nehru thought. He believed that India was due its own Enlightenment, like Europe's, to sweep out this disease of superstition and replace it with logic.

Today there is even a plea in the Indian constitution, announcing, 'It shall be the duty of every citizen of India to develop the scientific temper.'

And this is what allowed Vikram Sarabhai to get his ambitious space programme off the ground. Nehru had seen how the Soviet Union had made huge industrial strides by drawing up bold five-year plans for giant new factories, power stations and dams. He also knew that the US had made itself economically powerful on the back of innovative engineering. In a now-famous speech in 1960, Nehru announced that it was 'science alone that can solve the problems of hunger and poverty, insanitation and illiteracy, of superstition and deadening custom and tradition . . . The future belongs to science and to those who make friends with science.' He took this literally. He kept scientists like Sarabhai close to the government and used them to create the country's first physical research laboratory, nuclear power stations, hydroelectric dams and steel plants. Additionally he founded dozens of new universities and engineering colleges to train up the legions of graduates that would become the logical, rational, science-loving citizens that he had always wished for.

The change was slow. In the short term, Nehru's plans failed to pull millions of Indians out of poverty and into employment. When he died in 1964, the country was still poor. My dad graduated as a chemical engineer not long afterwards, from a university in north India. He worked for a while, for a chemicals plant in the eastern city of Kolkata, but in the end he left for Britain. And the same happened with thousands of other ambitious Indian scientists and engineers who emigrated that decade, staffing the hospitals, laboratories and universities of the West.

Change must have seemed hopeless back then.

But it would come.

Nehru had laid down the bones of a scientific infrastructure. There were powerful energy sources, well-stocked laboratories, efficient factories and a bespectacled army of trained researchers and engineers, many of whom would eventually return from their careers overseas. India was kitted out for battle in the coming technological age, even though Nehru never lived to see it.

'And then he said, at that time, he said you make me a report,' says Rao, absentmindedly shuffling through a stack of papers in a wooden box. In 1970 Vikram Sarabhai gave Rao the job of explaining exactly how the country's first satellite would work, while he himself went to work convincing the world that it was necessary in the first place. Rao tells me about a moment in Vienna, not long after the Russians had started their space programme, when Sarabhai was asked what a poor country like India thought it was doing by starting an expensive space programme. Echoing Nehru's speech, he famously replied, 'We do not have the fantasy of competing with the economically advanced nations in the exploration of the moon or the planets or manned space flight. But we are convinced that if we are to play a meaningful role nationally, and in the community of nations, we must be second to none in the application of advanced technologies to the real problems of man and society.' Making big scientific discoveries could be left to NASA, Sarabhai thought, India's space programme would instead solve some of the country's real problems – just as Nehru had wanted.

'The Indian space programme right from the very beginning was tuned to society. Without it we wouldn't have got any support,' says Rao. 'By and large it was to stabilise education and communication. We had only seven or eight television stations in the country then. So we made the decision that satellites are extremely important for our country.'

Still rustling through the box on his desk, he finally lifts out an old magazine article by Arthur C Clarke, about satellites. There is something scrawled in ink at the bottom. 'That is his signature there,' says Rao. 'He gave me this as a memento about four, five years ago.' In 1945 Clarke had come up with the radical suggestion that just three satellites, placed high enough in fixed positions relative to the Earth (known as a geostationary orbit), would be able to send and receive signals from the planet's entire surface. It meant that telephone connections could be made from one corner of the globe to any other place on Earth without the need for physical wires connecting the two.

For decades, Clarke's idea languished in science fiction. Technology and the demand for better communications caught up only much later. It's difficult to imagine it now, but in the sixties, explains Rao, 'when my wife wanted to talk to her father in India from where we lived in a little district in Boston, we had to wait for a whole three days and nights to get a call from Boston to there, and finally at the end of the third night we got a call through. And she takes up the phone and say hello, and her father say hello, and they get cut off.' The problem wasn't just bad lines; in India, particularly, the land mass was so vast that telephone connections were almost as rare as gold dust. It was around this time that the US and the Soviet Union started sending up their first experimental satellites, edging a little closer to Clarke's blueprint for a globe connected by three geostationary satellites.

India decided to follow suit, but it took many more years. In fact, Vikram Sarabhai died before the project was finished. Then in 1975 ISRO's design was approved and India's first satellite, named Aryabhata after the ancient mathematician, was built. Sent up in a Russian rocket, it wasn't a very useful machine at all, with a paltry resolution of a kilometre – only forests and seas could be made out. But as a test to find out whether Indian scientists could actually build a satellite at all, it proved they could. The satellite even earned a place on an Indian postage stamp.

The same year, Indian researchers used an American satellite to run an ambitious project called SITE, the Satellite Instructional Television Experiment, bringing television programmes in local languages to 2,400 villages across India. 'It was all educational programming, for six hours a day and every day, so farmers could see,' Rao reminisces with a broad smile. For farmers who had never used a phone, let alone seen television pictures, the programmes on health and agriculture marked the start of a communications revolution. Both the Aryabhata satellite and the SITE experiment were a success.

And this was the moment India's fortunes turned around.

Within a decade, ISRO launched its own Indian National Satellite System. It was a set of geostationary satellites exactly as Clarke had originally imagined, and it brought the same benefits to the rest of India as those first few villages had enjoyed back in 1975. More than just television, they brought communications, radio, mapping and weather prediction into the space age. It was an outcome beyond Rao's dreams. Little did he imagine that Indian scientists would be sending almost 50 satellites into orbit in the next few decades, or that this satellite network would eventually become the backbone of the country's huge telecoms industry.

Even a starry-eyed futurist like him could never have guessed that the early seeds of scientific effort planted by Prime Minister Nehru would bloom at exactly the same time. The laboratories created in the 1950s multiplied and the universities spawned entire armies of computer-savvy engineers. Today Indian engineers are designing software code for companies on the other side of the planet, connected to them only by satellite links. By 2015 these software companies are expected to be making as much as twelve billion dollars a year.

Nehru's sixty-year-old vision of a nation made stronger and wealthier by science may finally be coming true. 'The whole world is different,' Rao marvels. 'Now my son is there in Santa Barbara in California and in two minutes we get through to speak to each other. See, that's what I'm saying. The whole world has changed. Today we have more than hundreds of channels and communications is absolutely fantastic. All this came from those first satellites. Now we have something like almost half a billion phone connections in the country.' He pauses, and then concludes:

'With that first satellite, from then onwards we had created history.'

It's time for me to leave for the first part of my journey through India. I'm travelling north to the capital, New Delhi. But before I go, Rao wants to know the title of my book.

'*Geek Nation*,' I tell him.

'What nation?' he asks, squinting in confusion.

'Geek. G. E. E. K.'

He has never heard the word before. The problem is, I often have trouble explaining it. Internet definitions range from 'a person with an unusual or odd personality' to someone 'more comfortable with computers than with other persons'. Sometimes people think of a geek as the introvert in the corner of the party, or the comic-book collector who goes to *Star Trek* conventions, or the professor in thick glasses holed up in her laboratory. But to me, at least, geekiness is all about passion. It's about choosing science and technology or another intellectual pursuit, I tell Rao, and devoting your life to it. History's ultimate geeks are the men and women who sacrificed their lives on the altar of science, risking failure to pursue an obsession.

The sun streams in through the office window while his secretary brings us our final cups of tea. Rao looks frail. His friend Arthur C Clarke died about two years ago, many of his former colleagues have passed away too, and he himself should have retired more than a decade ago. In fact the kind of research that he used to do has long since been taken over by younger scientists and engineers. After all the change he has witnessed in the last sixty years, I can't imagine how Rao must feel to be missing out on the future, on space elevators, human settlements on Mars and extraterrestrial journeys of the kind he could only have dreamed about as a 20-year-old scientist, studying cosmic rays in a makeshift laboratory. India is only at the start, he muses, staring into the distance.

Walking outside, past the geometric flowerbeds in front of the Indian Space Research Organisation, I think about the geeky expedition that I'm about to embark on. About a year ago India's prime minister, Manmohan Singh, announced that he would double investment in scientific research from less than one per cent of national income, as it stands now, to more than two per cent. It's a landmark move, because this two per cent milestone is what generally separates the world's industrialised countries from the developing ones. The Japanese government, for example, spends more than three per cent of its income on research and development, in the US, the level stands at about 2.6 per cent, and in China, it's roughly two per cent. South Africa and Brazil, by contrast, spend about a per cent each. And so this may be the biggest sign yet that India is ready to join the big league. It was after this announcement that I decided to take a journey into India's future, to discover whether this nation can become a scientific superpower to rival the rest of the world. Or whether it is just hype.

In the coming year, I plan to meet young students at engineering colleges in the north, cutting-edge physicists in nuclear facilities in the south, geneticists in futuristic laboratories in the east and technology titans in the west. But, here in the Indian Space Research Organisation, I realise that my route also carries the footprints of the past, of countless scientists, engineers and leaders, like Rao, Sarabhai and Nehru, who helped turn this into a nation of geeks.

And perhaps it's also something from my past that has brought me here from London, to a country I first discovered as a child visiting my distant family and then after I left university, when I did a stint as a reporter in New Delhi. Since I was a kid, I've never understood why my dad is such a geek. Nor, in fact, why I turned out to be a geek too – as have many of my cousins. And it's not just us. Wherever in the world we live, Indians and people of Indian origin are famous for being swots, nerds, dweebs, boffins and dorks.

I want to know why.

CHAPTER 1

BRAIN GAMES

Sitting on one side of the trestle table is a ten-year-old boy with a dark-brown birthmark that circles one of his eyes like a panda, and on the other side is a middle-aged woman. Between them is a green and white plastic chessboard, closely watched by a bunch of wide-eyed schoolchildren.

She moves her queen to a square where he can easily take it with his knight. The boy strokes his chin. 'Why did you put this here?' he asks.

Realising her mistake, she drags it back.

I'm in an echoing school hall in an expensive neighbourhood in the heart of India's capital, New Delhi. And I'm here to find out what makes this country so apparently geeky. The morning traffic is roaring outside, with auto-rickshaws jostling for space between the scooters and cars, but inside are half a thousand people silently hunched around long benches, squeezing in a few precious moments of practice before the city's biggest ever cash-prize open chess tournament begins. It's so popular that the organisers have run out of seats. There are children with their elderly grandparents at the front, suited professionals and moustachioed taxi drivers leaning against the walls, and stragglers peering in through the doorway.

Indians love chess. The country is ranked fourth globally in

19

the sport (and chess players insist it *is* a sport), above the United States, which is ninth.

But there's something more unusual about this scene: generally, India is not a sporting nation. In fact, apart from cricket and a handful of other games, chess is a rare exception in a land that has one of the worst sporting records of anywhere in the world.

Take the Olympics, for example. In the history of the modern games, the United States has won 2,549 medals, Great Britain has won 737 and China, 429. Even the small eastern European nation of Belarus has taken home seventy-three. But you have to scroll almost to the bottom of the league tables to find India. It sits just above the desolate central-Asian republic of Mongolia and just below Slovakia. In the whole of the Olympics' history, India has won only twenty medals. Given the country's vast population, it's a mystery that has confused sportswriters for decades.

So a few years ago two US researchers, Anirudh Krishna from the Sanford Institute of Public Policy at Duke University and Eric Haglund from the Congressional Hunger Center, decided to investigate. A country of more than a billion people like India, they calculated, should have won 157 medals at the 2004 Olympic Games. But of course this fails to take into account that elite sports are expensive, ruling out millions of Indians who would never have a hope of becoming professional athletes. Wealth and size aren't the only things that determine Olympic success either; there's also the general level of education, people's health and how close they live to sporting facilities. So the researchers crunched the numbers again. Taking into account the myriad factors that determine sporting success, they came up with a far more conservative estimate.

India should have won around fourteen medals at the 2004 Olympic Games, they said.

In reality, though, the country won just one. In fact no other nation in their study had such a huge gap between its predicted medal count and the actual total.

'India did not, and does not, have a sporting culture,' the veteran Indian sports columnist, Rohit Brijnath, explains to me frankly. Today Brijnath works for the *Straits Times* newspaper in Singapore, after the Indian sports magazine he was writing for closed down. 'Personally, one of the things that I always felt as a sportswriter was a lack of drive among many athletes. I can't understand it. It is much better now, but earlier you only found that drive here and there, in exceptions like the great 400-metre runner of the 1960s, Milkha Singh, who used to boast that he trained so hard that he used to pee blood.

'My theory, and it's just mine,' he adds, 'is that we're better suited to hand-eye sports like shooting, or billiards, or archery. Or thinking sports, like chess.'

The cash-prize open chess tournament is about to begin in the school hall. A woman in an orange sari and wearing a red *bindi* walks towards the microphone on stage, flicking through a list of rules. 'Players, get ready to enjoy the game,' she announces, before the room transforms into a hive of shuffling pieces and clicking clocks. Each player is dreaming of glory; that they might one day earn the title of chess Grandmaster, shared by twenty-one other Indians, including the current world chess champion.

I turn to Radhe Shyam Tiwari, a 64-year-old international arbiter for the World Chess Association, for his opinion. He has been playing this game for forty years. Watching the players on the table next to us, he ponders the question for a while.

'Indians, well, basically they have a good liking for brainy games,' he announces at last, rolling each 'R' with his tongue.

'Yes, and we are good at brainy things,' he continues. 'Brain makes us supreme.'

'Our country's Olympics budget is more than a thousand million rupees, and those sportspeople only get one silver medal or something,' says Professor Vijay Singh, with a slight lisp. He pauses here before continuing (he likes to be dramatic). 'But we, *we* always get a gold medal.'

Standing at the front of the lecture theatre next to an old-fashioned blackboard, wearing stonewashed jeans and a flat cap, he's a former teacher who now runs the country's International Science Olympiad training centre. Although it sounds like it could have something to do with the Olympic Games, it's in fact the opposite. The Olympiads are a competition for high-school students, pitting their intellectual muscle against teams from around the world in a series of written tests and practical experiments in physics, chemistry, biology, mathematics and astronomy. Along with chess tournaments, they're some of India's most popular contests.

Today Singh is holding an 'exposure camp' for a hundred or so physics teachers, each of whom is hoping to get their students entered into the competition.

The questions on the projector screen in front of us are even tougher than we expected. Singh flashes up a slide carrying challenge number two from the International Physics Olympiad, which was held in Italy in the summer of 1999. It's called the Ampère versus Biot-Savart problem, he explains. We have to figure out which of two possible equations, one proposed by the French physicist André-Marie Ampère, and the other by physicists Jean-Baptiste Biot and Felix Savart, best describe the phenomenon of electromagnetism – why a wire with an electrical current running through it produces a magnetic field.

To our embarrassment, we struggle. Singh reveals that the summer this question came up, all five of the Indian contestants won a medal and one of them also earned a special prize for coming up with the best single answer (he not only worked out that Ampère was right, he gave *three* different right answers). The Indian teams always finish somewhere near the top in the

Olympiads, joined by a few other Asian countries, typically China and Taiwan, plus occasionally Iran.

These kids aren't exceptions, Singh tells us, they're the finest examples of an educational system that produces millions of the world's best science students every year. 'The Olympiad success story is the success story of every average science student in India,' he says, ending the lecture. He switches off the projector and spins round to face his audience with a smile. The teachers applaud excitedly and hang around for the next session, while Singh walks me to his narrow office next door.

'I do this because I enjoy doing it,' he tells me. His office is packed with physics books and test papers. There's a wrinkled red apple lying on a shelf. Around a hundred thousand school students turn up to examination centres every year in the hope of qualifying for one of his teams, and he brings the brightest few of them back to this centre to coach them before the big events.

'It all started in 1990 when I was visiting the United States, and I happened to meet two people who run the Olympiad programme there, and they happened to impress on me that I should send an Indian team. It took eight years to get it started after that,' he says, pulling out some photographs from a filing cabinet of former teams, lined up in proud, beaming rows with giant medals hanging from their skinny necks. A few are wearing custom-made blazers and holding Indian flags. Among the pictures is the physics Olympiad team who went to the United Kingdom in 2000, winning two gold medals and two bronze, ending up third overall in the world. Then there is Vietnam in 2008, where they took four gold medals, and Mexico in 2009, where they came home with four gold and one silver. A 19-year-old girl he once trained, who won a bronze medal in physics, has just had a paper on quantum physics published in an international science journal, he adds. Most students her age don't even read academic journals.

These days the Olympiads get as much coverage in India's

national press as the Olympics – perhaps even more. As far as Singh is concerned, this is because Indians view a gold medal in science as more valuable than a gold medal in cycling or sailing. 'There is this mindset among Indians that knowledge is good and science is even better. It's a cultural thing. And sport, they think of sport as a recreational activity not to be taken seriously,' he says.

Like the old chess player told me, in India brain is supreme.

When India entered its first Olympiads and returned with a clutch of medals, Singh wasn't even surprised. 'And I'll tell you why I wasn't surprised,' he says. 'One of the people who run the Olympiad programme in the US, Robert Resnick, he gave me a problem. Now, Resnick has written one of the most important textbooks on physics. I was teaching back then and I knew the students could tackle this problem. We used his textbook, and I knew they could tackle this problem.'

At the heart of India's obsession with braininess is the school system. And the reason the Indian school system demands so much of its students is partly because of the intense competition for college places. Last year 472,000 teenagers sat the national engineering college entrance exams, and only 10,000 of them won a place. This two per cent acceptance rate is a fraction of world-class universities; lower than even Harvard and Oxford. Millions of Indian teenagers willingly relinquish their entire childhoods to exam revision, tutoring and studying sample papers.

At the top of the academic league table, where Singh's students are found, education isn't just education; it's one enormous talent contest, like the US National Spelling Bee (which, incidentally, almost always has some Indian-origin contestants), but larger. And the kid with the top grade becomes an automatic national celebrity. One year, newspapers reported, a shy 18-year-old who got the highest mark in the national engineering entrance exams was so shocked by the clamouring queues of students and neighbours desperate to meet him that he ran away from home. Refusing to lift his eyes from a book, the frightened boy told journalists, 'My house is no longer the normal place I lived in all these years.'

Singh has seen lots of teenagers in the same boat. He shows me photographs of the Olympiad teams once again, this time pointing out other students who got remarkably high marks in their college entrance exams. Every picture has at least one prodigy. 'He was ranked number one,' he says, pointing to a tall, gangly boy called Nitin Jain, part of the 2008 Astronomy and Astrophysics Olympiad team.

'I think I've heard of him,' I say.

'Yes, you might have. He's in the papers a lot. He's an interesting child.'

From what I've read, Nitin Jain is a wunderkind. Not only did he beat almost half a million teenagers in his exams, he has also brought home medals in three Olympiads, from Mexico, Indonesia and Iran. Since his success, he has appeared on television and done countless newspaper interviews. I recall hearing that he's even writing a book about the reasons behind his phenomenal success. If I'm ever going to understand the prototypical Indian nerd, I have to meet him.

Mathura Road, which links New Delhi to the town of Faridabad where Nitin lives with his parents, is a wide and dusty dual carriageway lined with industrial warehouses and office blocks. While New Delhi is a maze of curved streets sprinkled with ancient architecture and green parks, Faridabad is more of a work in progress. The landscape is uniformly beige. Small houses and tall apartment blocks have been constructed in the middle of dry fields, connected by dirt roads with crudely painted trucks rolling in between, carrying bricks and cement to build even more housing. It's a pop-up suburb, and the kind that encircles many Indian cities nowadays, designed to house the hordes of aspiring villagers who want to move to the cities and find jobs and better schools. Flanking Mathura Road are three giant billboards for coaching centres, claiming to help students earn higher marks in their exams by intensively tutoring them for a fee.

Nitin's house is like the others here: new and small. The bare, dark living room has one small glass cabinet full of heavy trophies and on a bench in one corner is a laptop that Nitin won at an Olympiad. His parents tell me proudly that he will be the family's first engineering graduate. His father works part-time in a factory and his mother, a beautiful woman in a sari and pink lipstick, is a housewife. Neither speaks more than a few words of English so we stick to Hindi.

Before Nitin arrives, I catch them glancing at each other and then sharing a whispered conversation.

'Please don't record the interview,' they ask me, quietly. 'Please.'

At that moment, Nitin walks in, barefoot, toweringly tall and thin. He is so shy that he hardly looks at me. And he has a debilitating stutter. He repeats some words four or five times before he can finish a sentence, and the sentences are frustratingly short. If I have to record the interview, his parents ask me, then can I please erase it after it's transcribed so people won't hear him?

I promise to erase it.

Throughout, they won't leave the room, bringing apple slices, pieces of orange covered in spices and diamond-shaped crackers.

Slowly, Nitin opens up. Having done sample papers beforehand, he says, he always expected to come somewhere near the top in the country in the college entrance exams. At first I assume that he's boasting, which is surprising for a kid who's so shy. But I come to realise that he's simply stating a fact as he sees it. In the final test, he got 424 marks out of 480, the next highest was 417, and the average was 100. It wasn't a close contest. He was always going to beat most of the students in his year.

When the results arrived, he adds, his parents threw a party for him that attracted such widespread news coverage that he's been spotted by strangers in the streets.

But it seems to me that he doesn't enjoy all the attention. 'Most of my time I spend on Internet,' Nitin continues, painfully slowly and without lifting his eyes. The students at his college love multi-player online games, eating up their evenings with *World of Warcraft* and *Counter-Strike*. But he prefers *Grand Theft Auto*, he tells me. He likes reading *The Lord of the Rings*. And he plays the occasional game of table tennis with his friends. But this is about all he tells me. I spend more than two hours in his home, but struggle to draw any more out of him.

'I'm thinking of studying physics. Physics is very interesting and it's fun to do it,' he says, finally.

'So why do you like physics?' I ask.

'I don't know about it for sure. I don't have any idea about what I'll do after graduation. I don't have any idea,' he says, looking at his father out of the corner of his eye. 'I am still not sure about all these things. I mean I am studying engineering now, after completing it, I will see what to do. I am not thinking now, I am still not sure about it.'

'You have no idea then?' I ask, a little surprised.

He looks down. 'No.'

I'm disappointed. I was expecting a curious geek. I had assumed Nitin would love science so much that he would scarcely be able to contain his enthusiasm for it. Stutter and shyness aside, he can't even tell me why he chose his degree course or which bit of it he enjoys most. This is strange behaviour for a wunderkind. Engineering students tend to be tinkerers and hackers (I built model rockets when I was at school, for example), but he tells me that he's never even spent a minute of his spare time playing with gadgets or taking apart household appliances. He's not geeky; instead he looks overworked.

'He has written this book, let me show you' his father interrupts in Hindi, enthusiastically handing me a bunch of stapled papers, filled with photocopied newspaper clippings and photographs. He has invited over some neighbours so they can meet me and find out what I think about it. They wander in, in file, shaking my hand before they sit down.

'What do you think?' he asks, beaming with pride. They all look at me expectantly.

Flicking through the book, I see long essays about how Nitin does his work, what apparently motivated him, his family life. There are flowery paragraphs about how his devoutness to his Hindu faith helped him to win Olympiad gold medals and how he owes so much of his success to his kind and loving parents. 'It's nice,' I tell his father (but it doesn't sound like him, I think to myself).

'I want him to be the top of the world,' he replies, smiling the whole time. His wife and their neighbours smile too and nod in agreement.

'Top what?' I ask.

'Top scientist,' he replies.

Nitin's parents and neighbours leave the room for a few moments and I grab the opportunity to ask him a few questions in private. He relaxes and his stutter becomes less pronounced. Reaching for a slice of apple from the table, he tells me that he had pretty much nothing to do with the book. He says his parents wrote most of it.

He isn't even very devout, he continues. He goes to the temple only once every two or three months, for festivals.

So before I leave, I ask Nitin the real secret behind his phenomenally high grades. For two years before taking his college entrance exam, he tells me, he devoted four hours each weekday evening to studying and spent every weekend being coached by tutors. 'The secret is hard work,' he says.

Indian scientists and engineers have a universal reputation for being quiet, shy, hardworking types, without much creativity. And I can see now how there might be some truth to the stereotype. Vijay Singh from the Olympiad training centre tells me that he was training Nitin for an Olympiad just a few weeks earlier. The reason he does so well, he says, is that he's a very bright kid who simply happens to memorise things a lot faster than most.

And in India at least, a good memory is all a student needs. There are two reasons for this. Education is relentlessly hierarchical, which in the past has made it difficult to ask questions or challenge authority. 'It comes from some ancient times maybe, that knowledge is to be revered and respected,' Singh suggests. 'I cannot speculate on that, but it is obvious if you travel on the train or something. People tend to respect you if they know you are a teacher. That culture is respected and it still persists now, that attitude.' This may be why the linchpin of the Indian education system remains old-fashioned learning by rote. This technique, used across the continent, means that Asian kids can't be matched when it comes to doing quick calculations in their heads or naming capital cities.

At the same time, many Indian teenagers dream of careers in medicine or IT. A 2009 survey by the auditors Ernst and Young, for example, found that India has the largest pool of scientists and engineers in the world. Two million students graduate from almost 400 Indian universities every year, including 600,000 new engineers. And since at least a third of the population lives below the poverty line, for most of them, this is a ticket to a good salary and a stable future for their families.

The result of these twin pressures is a tough, but purely theoretical, exam system. A kid with a good memory can get a decent grade if they just put in enough hours of practice. And in the last decade, the pressure to cram their minds with ever more information has forced millions of children into supplementary classes in the evenings and at weekends. Coaching centres litter most Indian towns, with tutors advertising everywhere from newspapers to posters in the streets. A typical advert will carry a row of photographs of the top students, with their percentage exam scores in big black letters underneath. In fact, in Kota, in the state of Rajasthan, there are some coaching institutes to help students get into other, higher-rated coaching institutes. The industry does such brisk business that, in 2009, one particularly successful tutor in the small Indian city of Patna got death threats from rival coaches after he managed to earn a 100 per cent success rate for two years in a row. 'I have security guards now,' he told reporters.

'There is intense competition to get into the good courses, post higher secondary school education,' explains Singh. 'When it comes to universities, you'd like to go to a very good university, or a very good engineering college or a very good medical school. That's it. So students tend to prepare for it. Now parents are also fully behind them, so parent support, student motivation and in addition very good coaching classes, which are like a parallel education system in India, they all combine to give a pool of students to us who do well in exams.'

The problem is that the best scientists and engineers are not always the best at passing exams. Albert Einstein, for instance, made his observations on relativity in his spare time while working as a lowly patent clerk. In science, Vijay Singh tells me, you need to be able to think independently and outside the box. Great research and inventions don't come from memorising the multiplication tables.

'But things are changing,' he adds. 'There are top-notch students in India who are very creative. If you want to see, you should go to one of the engineering colleges, especially the IITs.'

Stepping off a noisy highway in the affluent district of Hauz Khas, in New Delhi, I think that at first glance the campus of the city's Indian Institute of Technology is actually the grounds of a luxury sports club. Peacocks strut across lush, irrigated lawns. And a long sweeping driveway invites me towards a pink-tinged multistorey monolith ringed by squash courts and an amphitheatre. This is where Nitin Jain is studying, along with 800 of India's cleverest young scientists and engineers. And I've been told that it's a real geek factory. So if India really is becoming a scientific superpower, this is the next place to look for shoots of innovation and creativity.

Founded in the 1950s and 1960s by Jawaharlal Nehru, today there are sixteen Indian Institutes of Technology, known as IITs, dotted around the country, with plans to open another eight, including the first international IIT college in Qatar. They were originally built as part of Nehru's master plan to train up an army of innovative young engineers, who would be the country's first generation of technocrats, researchers and inventors.

And to a large extent, they've accomplished that mission. The IIT global alumni association claims that its graduates have founded at least a fifth of all the startups in Silicon Valley in California in the last two decades. And Bill Gates, the founder of Microsoft, which itself employs dozens of former IIT students, once said, 'It's hard to think of anything like IIT anywhere in the world. It is a very unique institution.' The brand is so legendary that in 2003 it made it into a famous Dilbert comic strip. In the sketch, an Indian office worker tells Dilbert, 'Luckily I'm an IIT graduate, mentally superior to most people on earth, so I finished the project myself'.

In the grounds, I spot one student wearing a college T-shirt, emblazoned with 'I am what you dream to be.' I've also heard that among the myriad items of institute merchandise you can buy online is a pair of IIT Delhi boxer shorts, for 600 rupees, which is about thirteen dollars.

Walking through the maze of dusty, grey corridors, I reach the dean's office. Professor Santanu Chaudhury is an IIT alumnus himself; he graduated in 1979. There are two middle-aged men sitting in the reception outside his room. One is his secretary and the other's job, from what I can tell in the twenty minutes that I am waiting here, is to open the doors and fetch the occasional cup of tea (Indian offices are always stuffed with peons like this one – a reminder of how rigidly hierarchical society still is, reflected in the education system). I ask him if he might allow me to spend a week here, mixing with the students. It's the only way I'll get to know what the place is like, I explain.

'You know that end-of-semester exams are in a week?' he asks.

'No. But don't worry, I won't get in the way.'

'Make sure you don't.'

He asks his secretary to fetch me a timetable and a map. My first class, on turbomachinery, is at eight o'clock this morning. This is one of the toughest courses, a student tells me, and it's full of only hardcore geeks.

I'm already late and the lecture theatre is full. There are hard wooden benches in long rows, illuminated by slits of sunlight through the dirty windows. A slide on the projector explains the first step in designing a small centrifugal compressor, of the kind often used by small aircraft to pump and squeeze fuel gases into the jet engine, maximising the energy that comes out of it (the lecture notes enthusiastically describe it as 'A Compact Life Generating Device!!!').

'Now we need compressors because, suppose I'm going to build a commercial aircraft, I can't give it ten engines, can I?'

says the professor. 'That would look stupid.' The nerdy students
in the front row let out a ripple of laughter.

Sitting in front of me is Prerit Jain, a 22-year-old mechanical
engineering student, who tells me that he's at the top of his final-
year class with a grade point average of 9.48 out of ten. He's the
only person here brave enough to ask a question throughout the
entire lecture. It's a tough course, he tells me, but all the courses
are tough at the IIT. Like most of the students here, Jain takes
between twenty-six and thirty-five hours of taught classes, lectures,
workshops and tutorials every week. On top of this, in a light
week, he works for an extra ten hours, and in a heavy week (which
is most weeks) as many as thirty-five. 'The effort you have to go
through to get here really toughens you up. I mean, for a fresh
student, their first exam is less than a month after they arrive,' he
whispers to me across his desk when the class ends.

Afterwards, Jain and I walk to the main hall where students
are hanging out between lectures. Although the campus is
beautiful from a distance, parts of it are a sprawling mess close
up. With the morning mist low on the ground, the dark, grimy
courtyard outside looks like a cemetery. 'When I first came here,
I was disappointed,' he admits, throwing his bag onto a table.
'If a place is so hyped up, you expect it to be a dreamland. Having
travelled through Europe and seen what it is like there, I have
to say – it's a strong word – but this place is filthy.'

Despite its reputation, IIT Delhi doesn't have a slick European-
style university campus. Instead, many lecture rooms are lined
with rickety benches, with respite from the heat coming only
from a row of ceiling fans. The Indian government spends under
200 million dollars a year on all of the Indian Institutes of
Technology, equivalent to just eight per cent of the annual budget
of the Massachusetts Institute of Technology in the US. Teachers
always complain about the low wages and students have been
lobbying to have air conditioners fitted in the classrooms for
years.

Even so, Jain turned down a scholarship at the prestigious Nanyang Technological University in Singapore to be here. 'I came here because of the aura of the IIT,' he says. However dismal the surroundings, the brains are hard to beat.

For all the students, work comes first. I was an engineering student once but I've never seen an atmosphere like it. Hardly anyone plays sports, social mixers are unheard of and a lot of the boys still have trouble talking to girls, says a girl who joins us for a chat in the main hall. 'The party culture is not close to even what you would expect,' agrees Jain. Between the lecture room and the grey main hall are huddles of students, cramming for the exams in a week's time. One has the hood of his yellow sweater up over his head while a group of girls sit in a stairwell nearby, desperately comparing notes.

Later I meet Nishant Ranka, a 22-year-old final-year electrical engineering student. 'When I went to the UK and saw people shut their shops at five o'clock, I couldn't understand it,' he says. 'Here they are open late into the night. Indians work a lot, lot harder than Europeans and Americans.' He tells me that before he arrived at the IIT in Delhi, he was coached for three hours a day, as well as the nine hours a day he would study anyway, to make sure that he passed the exams. He occasionally allowed himself half an hour of cartoons in the evening, and caught a movie after an exam was over, but otherwise his life revolved around the crucial test.

By the time Ranka arrived at college, he was burned out. 'I spent one semester after I got here in my room, just playing *Age of Empire* on my computer. It was only when my grade point average slipped to 7.2 that I gave it up. I had to let it go because I thought I was wasting my life,' he says. He reminds me of Nitin Jain.

I rush to my next class, in a lecture theatre upstairs. It's one of the few humanities courses taught at the institute, called 'Macro perspectives on science and technology in humans'. It is not popular. In fact I'm the only person in the gloomy room at the top of a flight of grey stone steps for the first ten minutes.

'I'm going to start anyway, whether people show up or not,' Professor Ambuj Sagar announces, in an Indian accent with a strong American twang, just before a few latecomers arrive. The topic of the lecture is fascinating – the social reasons that automobiles went from small invention to mass-market marvel in the twentieth century – but almost all of the students stare at him blankly, uninterested. If it won't help them pass their engineering exams, they don't want to know. None of them are willing to ask questions, and even when he picks people out to speak to the class, they look uncomfortable. Returning the latest essay papers, he tells them, 'I was a little bit disappointed that having harped on about innovation systems, the answers were not that impressive.'

By the end, he seems exasperated. 'An innovation system is not an object, it's what's in our minds,' he implores.

I meet up with him after the lecture, curious to find out how he thinks this college has changed since it was founded. Sagar studied engineering here at IIT Delhi in the 1970s. Like many graduates back then, and many even now, he left to work in the United States. After decades as a researcher at both the Massachusetts Institute of Technology and Harvard University, though, he returned two years ago to help improve educational policy in India.

'Students just don't care anymore,' he complains. 'What unfortunately globalisation has done is brought to India globalised opportunities and salaries, and that is the incentive that students are looking at. The salary differential between the bottom and top job in my time used to be a factor of two or three, and now it's something like ten or even twenty.'

The highest salary offered to an IIT engineering graduate last year was close to four million rupees, about $80,000. In comparison, the average Indian high-school teacher is paid around 100,000 rupees a year. This change has only happened recently, all because of a boom in India's IT industry that has sent the demand for computer engineers soaring. And the problem has become a vicious spiral.

Teenagers like Sagar used to come to institutes like the IIT out of a geeky love of brainy games. They did so well that the Indian IT industry grew, attracting more kids into engineering. These graduates turned India into the technological nation that it is today. Computer engineers have overtaken doctors and lawyers in India's social hierarchy (in fact when I tell people that I'm now a journalist, they assume I must have failed my engineering degree). These days, engineers are the most eligible people in newspaper matrimonial adverts. So much so, in fact, that male and female engineers often end up hitched to each other. But in a nation still split by wealth, millions of young people who are now desperate for lucrative technology jobs are finding that there just aren't enough college places or jobs for them all.

So unlike in Sagar's time, when teenagers applied to the IITs because they were good at science and engineering, students nowadays just want plum jobs – even if this means dedicating all their spare hours to cramming for exams.

Estimates suggest that at least half of all the IIT students have been through coaching. The problem has become so severe that, one student tells me, there's even a Bollywood movie out now that highlights the problem. *3 Idiots*, he tells me, is about a creative genius who finds he is an exception in an engineering college in which all the other students are hardworking grade-chasers. One of the characters is so broken down by the system that he hangs himself.

'There is only one reason the students come here. If you get into the IIT and then get a job when you leave, you're likely to earn more than your father does,' says Nishant Ranka, the electrical engineering student I met earlier. Graduates here are so sought after that the top four in this year's batch haven't even bothered to sit for interviews in India, comfortable in the certainty that they will secure high-paid jobs in one of the dozens of multinational companies desperate to hire clever Indian nerds.

On Monday morning I pay another visit to the dean of undergraduate studies, Professor Chaudhury, this time to find out what this all means for the future. I want to know how the pressure for cramming for exams, and the lack of creativity, has affected the college.

The large desk in his office is strewn with files and papers and there is a queue of students waiting to speak to him. He can only spare me ten minutes, so he gets straight to the point. 'I'd like to see less coaching,' he says. In recent years he explains that the coaching system has turned students into little more than robots, trained to beat the exam system and unable to think for themselves. 'I don't know if the coaching centres skew the kind of people we get, but in many cases for two years they are under tremendous pressure, which is not good for the personal development as teenagers. All-round development is essential. Some children are even away from their families for two years,' the dean says.

Burnt out before they even arrive, hardly any IIT students stay on to do research or further degrees. Only a tiny proportion of them are interested in careers in a laboratory.

And the pressure is starting to take a more dangerous toll. A 2005 survey of Delhi schools by the Vidyasagar Institute of Mental Health and Neurosciences discovered that 40 per cent of students felt overwhelmed by exams. Meanwhile in the nearby city of

Lucknow, psychiatrists have reported that some children suffer from hysterical psychosis brought on by a phobia of exams. According to the most recent figures from the National Crime Records Bureau, more than 2,000 students committed suicide across India in 2008 because they failed their exams – a ten per cent rise on the year before. In ten days in 2010, fifteen students killed themselves in the western Indian state of Maharashtra alone, with rumours circulating in the press that exam stress was to blame for many – if not all – of the deaths.

My week at Delhi's Indian Institute of Technology is drawing to an end. I had hoped to find a geek's paradise here, but I haven't. Instead of being the hothouses of intellectual curiosity and innovation that they promised to be when India created these grand institutions fifty years ago, the IITs seem to have developed a culture of getting the grades and getting out. Students are literally working themselves to death.

Jawaharlal Nehru had dreamed of building a nation with a scientific temper, with a love of logic and brainy things. He had wanted a rational, trained workforce that would slowly lift the country's fortunes. But I get the sense that the reality is now more of a technological dystopia. I haven't yet seen any signs of creativity and passion – those qualities that lead to scientific discoveries and exciting new inventions. The country may have a booming IT industry, which has attracted millions more young Indians into science and engineering, but the pressure to get one of these lucky jobs is burning out their brain circuits, disabling their imaginations and driving hundreds to suicide.

They aren't geeks, I think to myself. They're more like drones.

The only way to understand what went wrong is to take a step back. It's difficult to know what drives a country as poor as India until you leave the city. So I'm travelling to the poorer district of Mustafabad, on the eastern outskirts of New Delhi. I pass a giant water pipe that stretches for kilometres, surrounded by heaps of household trash being nuzzled by pigs. The stench of sewage carries through the town. Now blocking my taxi on the dirt road is a mountain of scuffed computer casings, stacked one on top of the other like cheap plastic chairs.

After a few minutes, a man with a handcart comes along and hauls them away.

The street I'm in is well known in Delhi as an informal processing centre for electronic trash. This is where technology comes to die. Every building has been turned into a tiny workshop, loaded floor to ceiling with gadgets. One has hundreds of television monitors, another is scattered with circuit boards ripped out of computer mainframes, while a few have stacks of air conditioning units and refrigerators.

Children sit together in busy swarms, separating the useless pieces of metal, plastic and wire from the ones they might be able to sell. Their hands are covered in scratches and sores from where they've tried to rip apart hardware with their hands, and a few have also developed coughs and headaches after breathing in the toxic fumes from burning plastic parts.

There are dumps like this all over India. The country imports around 50,000 tonnes of electronic waste a year from richer countries, and in New Delhi alone, around 30,000 people sort it, picking through mountains of batteries and plastic screens at risk to their lives. Sorting through this stuff is expensive in the West, but the kids in Mustafabad do it for nothing, selling the old parts that they can salvage for about three dollars a day.

These children are at one end of India's technological labour chain.

And at the other extreme are India's computer engineers.

They may work in offices instead of crouching in the streets, but there are hundreds of thousands of them, just like the children sorting electronic waste in Mustafabad, and many of them are unfulfilled and overworked. Sometimes, these cheap IT workers are called 'cyber coolies', like sweatshop workers or technical busboys and busgirls. Often, their work doesn't need much imagination, just mind-numbing days remotely fixing software and processing billions of lines of code for companies in Europe and the United States.

Every year, many of India's engineering graduates end up in jobs like these. Some of them head to the offices of Tata Consultancy Services, one of the country's largest technology companies, founded in 1968. Here, I hope to understand why the country's mammoth IT industry has failed to inspire scientific interest and innovation among students like those at the IIT in Delhi.

'If you ask an IT guy about general knowledge, he will not even know,' admits Thomas Simon, a tall man with a thin moustache, responsible for hiring employees at Tata Consultancy Services. Simon joined the firm in 1990 after working for a company that made Coke cans. Every year he sends out recruiters to the top engineering colleges across India, including all fifteen of the Indian Institutes of Technology. 'We can do a thousand job offers in a single day. In fact we're in the *Limca Book of Records*,' he says. More than two million Indians work in the software export industry and Tata Consultancy Services hires 140,000 of them.

'An engineer here, he'll give you the best code to run the best software for the number one company on Wall Street,' he says. 'He is completely ensconced in his customer, in the solution, which he needs to provide. He may work fourteen hours a day to get that answer. And that's the alpha and omega of things.' This is one reason that India's technology companies are renowned for their quality and speed, creating the IT boom that India is now famous for.

The phenomenal growth in the industry happened in just a couple of decades. It started around 1995 – that was when firms like Tata Consultancy Services got a big break thanks to a computer bug that emerged just before the start of the new millennium.

Many decades ago, Western computer designers had come up with the economical idea of not wasting lines of code by, instead of typing the entire year into software programs, just typing in the final two digits. So instead of 1988, they put 88. And instead of 1999, they put 99. Sparing these two digits, millions of times over, saved countless bytes in precious computing memory and millions of dollars in expensive electronic storage. What they had failed to account for was that, by the time the year 2000 rolled around, computer calendars would switch to 00 and as far as any mainframe was concerned it could just as easily be the year 1900 as the year 2000. The problem became famously known as the millennium bug, or the Y2K bug. Governments and businesses around the world went into panic that this temporal crisis might erase bank records, force planes out of the sky and disable nuclear power stations. They needed someone to fix their software.

And that's when Indian software companies stepped in. Thanks to years of investment in universities and laboratories, partly because of Nehru's early commitment to science, it was around this time that India happened to have a glut of overqualified, underemployed engineers.

Rajdeep Sahrawat, now the vice-president of the National Association of Software Service Companies, a trade body for Indian computer businesses, witnessed the whole thing. 'IT managers in the West, they quickly realised that there were billions of lines of code, which somebody had to check. And they didn't have people to do it,' he says. On the black marble desk in Sahrawat's office is a row of miniature aeroplanes that he collected on his travels around the world to meet the companies who needed India's help. They included airlines, pension firms and banks.

Sahrawat had himself been a student at one of the government's elite colleges, a National Institute of Technology, graduating in 1984. Computers were so rare back then that he completed a degree in computer science without touching a keyboard – everything he learned was theory. But by 1995, he and the rest of the world realised that things had changed. Indian computer scientists were as good as those found anywhere else. He tells me that at least three quarters of the people working at Tata Consultancy Services in the 1990s were graduate engineers, about a fifth had postgraduate degrees and hundreds also had PhDs. It was one of the most highly educated workforces anywhere in the world.

India had two other advantages over countries that might have taken the millennium bug work instead. Firstly, Indian engineers all spoke English – a result of being under direct British rule for eighty-nine years – and secondly, they were cheap. 'You know, people say that size doesn't matter. But when you have half a million engineers available at an affordable wage, well, that changes everything,' says Sahrawat.

But the work they were asked to do wasn't particularly creative. 'It was, you know, grunt work. It was not sexy,' he says. 'Indian IT companies were smart enough to figure out that this was not work which had to be done in a project mode. It had to be done almost like a factory, an assembly line.'

India's drawback was that it didn't have the same resources as companies in Europe and the US. It could take years to import a computer into India and, even into the early 1990s, companies like Tata Consultancy Services were using spools of film, thousands of feet long, to record their data. They had to send heavy computer tapes straight to their customers because they did not yet have Internet connections fast enough to communicate with them any other way. One Tata Consultancy Services employee tells me that when Microsoft released its first Windows operating system, it wouldn't run on his computer because it was too slow. Many Indian software designers were stuck with a programming language invented in 1960, called common business oriented language, or COBOL. It had been used decades earlier in many of the earliest computer applications (for example, many of the first airline electronic-booking systems were written in COBOL).

But by a lucky quirk of fate, when fears about the millennium bug began to circulate, this weakness also worked in India's favour. Many of the software programs that were potential victims of the millennium bug had been built in the 1960s and 1970s using COBOL. In 1997, there were still more than 200 billion lines of working code written in COBOL, much of it in banks, government departments and airlines in the United States. Companies found themselves desperately calling back old programmers from retirement because they were the only ones who still knew how to use COBOL, before they realised that Indian firms like Tata Consultancy Services could do the work instead, both faster and for far less money.

'Y2K proved to the world that Indian companies could deliver world-class quality at an affordable price, remotely. This notion that India is just snake charmers and elephants started going away,' says Sahrawat.

By the time billions of lines of code had been checked and the year 2000 had passed by peacefully, the Indian IT industry had cemented its place in the global-technology landscape. Tata Consultancy Services was making around a quarter of a billion dollars in 1996, but by 2003, its turnover was more than a billion. Meanwhile, hundreds more IT companies sprang up in technological hubs like New Delhi, Hyderabad and Bengaluru. They were the places the world would go for a reliable, workaday engineer.

And this is where India's geeky engineering graduates ended up, selling their skills to the best bidder.

After 2000, the IT industry became a black hole for dronelike programmers, absorbing thousands of graduates who might otherwise have become laboratory researchers or inventors. Some of the time, their jobs were as basic as making sure that software wasn't broken. 'From 2002 to 2005 you started seeing these large contracts being given to Indian IT companies,' says Sahrawat. 'They were told, here is a bunch of a hundred applications, now you just maintain that. If there's a defect, you fix it. If nothing goes wrong, just keep staring at the screen.'

'You know, it's not about sweatshops anymore,' says 59-year-old Ravindra Shah, the chief compliance officer at Tata Consultancy Services. He's a small, round man wearing a stripy shirt and he's been with the company since 1975.

'Really?' I'm doubtful.

He lifts himself out of his chair and walks me down the corridor to meet the person in charge of innovation.

His name is Ananth Krishnan and he's a bit of a star in the technology community. What strikes me first is his unusual face. The crown of his head is balding and slightly larger than the rest, giving the impression that there's a disproportionately big brain inside his skull. Perhaps there is. When Krishnan graduated from the Indian Institute of Technology in Delhi in 1985, he was one of its most celebrated geeks. 'I stopped playing cricket after I went to the IIT, even though I was fairly good at it. I settled for academics and quizzing. We were all geeks,' he says in a voice that reminds me of Kermit the Frog from *The Muppet Show*.

Krishnan won the Brain of IIT title one year, and the Cranium Cup twice.

A couple of years later he joined Tata Consultancy Services. His first job was to design the software systems for a major

Scottish pension fund. 'A lot of Scottish pensioners probably owe their pensions to me,' he says.

'So what is the Indian IT industry bringing to the world that's different?' I ask him.

'What the Indian IT industry has done over the last twenty-five years has been very conventional, you and I know that,' he admits, almost apologetically. 'We've all been fairly conventional, on the incremental path. That is the hallmark of a stable industry, which is all right. We brought in quality, we brought in scale, and all those good things. But in the last two or three years the opportunity we've been trying to live up to is our ability to think differently.'

There's a pause. I wait for some examples.

'We're just going to be launching a scale pilot of our Internet-on-TV device, which is coming out early next month,' he continues, hesitantly. 'It's already finished a trial with 2,000 users and now we're scaling it up to 25,000 users. And we're experimenting with social networks.'

'Is that it?' I ask, looking at him. Both of us know full well that neither of these things is radically new or inventive. US companies have been developing Internet services for televisions for at least three years already. And social networking sites have been around for even longer than that.

He smiles awkwardly, struggling to think of other examples.

The uncomfortable truth that Krishnan is wrestling with is that Tata Consultancy Services doesn't put nearly as much effort into basic scientific research – the font of new inventions – as foreign companies do. Europe, the US and Japan remain light years ahead of India. Microsoft spends about nine billion dollars on research and development in a year, the computer-chip maker Intel, spends almost six billion dollars, and Samsung is aiming to triple spending to more than 22 billion dollars. Meanwhile Tata Consultancy Services spends only one third of a per cent of the money it makes on research, amounting to only about 16 million dollars. It's a strategy that depends on other firms coming up with inventions so they can piggyback off them.

'The Indian IT industry has been great when it comes to the application of IT, but we still don't create core IT,' Rajdeep Sahrawat from the National Association of Software Service Companies told me when we spoke.

That seems to be true, but as I'm wandering around the Tata Consultancy Services buildings, I come across rows of rooms stuffed with young researchers. One is a replica of the inside of an aircraft cabin, another is a fully functioning shop of the future, while more are sealed behind electronic doors. If they're not coming up with inventions or filing patents, then what are these people doing?

Ananth Krishnan has a simple answer: they're making things cheaper, he tells me. Tata Consultancy Services has built vast workshops in which engineers do nothing but shave incremental dollars off the price of popular gadgets and appliances by redesigning them. 'Our big advantage is that we think frugally,' he says, in a tone that reminds me of a door-to-door salesman.

'The challenge is to use today's technology but do it at a totally different price point. That's what we are trying to do first,' he continues. And making existing things cheaper is something that India has always been good at, ever since it started carrying out affordable software maintenance in the runup to the millennium bug. This kind of innovation may not be particularly groundbreaking, but cash-strapped Indians really appreciate it. They don't worry about this in America or Europe, because consumers there can afford to buy whatever Silicon Valley produces. In India, gadgets have to be far cheaper. Just recently, Krishnan tells me excitedly, one of his teams invented a water filter that costs only $24. It uses rice husks and doesn't need electricity. It was also the Tata group of companies that came up with the famous Nano, the cheapest car in the world, which sells for about two thousand dollars.

It reminds me of my encounter with Udupi Ramachandra Rao at the Indian Space Organisation, at the start of my journey. He had told me that India's space programme had started because of a need to create technologies that would benefit the poor. They started off mimicking American rockets and Soviet satellites, but now they're doing original work in their own right. And there's certainly something to be said for this approach to science and engineering. After all, this is how Japan launched its electronics industry in the 1970s. The island nation became a scientific powerhouse by undercutting American manufacturers on imported technologies like automobiles, computer hardware and consumer gadgets, while maintaining great quality and an educated workforce.

Michael Cusumano and Chris Kemerer, a pair of researchers from the Massachusetts Institute of Technology in the US, carried out a study in 1990 that found that, 'to outside observers, Japanese engineers . . . seemed to lack creativity and the ability to invent new or sophisticated products.' They were right. To begin with, Japanese innovations did tend to focus on management practices and how factories were organised, rather than fresh science and new products. But as they got bigger, the Japanese developed their own cadre of inventors. And they ended up doing research and making products that the world had never seen. Nintendo and Sega created bestselling videogames consoles, and after several more years, Japan produced the world's first humanoid robot. Today, it's a scientific research leader. This strategy of reducing costs first before moving onto innovation is where India's technology industry seems to be headed next. And just as Japan took the lead in hardware, India may become the leader in software.

But if there are sparks of geeky innovation here that might lead to this kind of scientific revolution, I haven't spotted them yet. All I can see are drones.

CHAPTER 2

ELECTRONICS CITY

I'm not even paying attention anymore.

'You are aware of the Gen X and the Gen Y?' Thomas Simon asks me.

'What?' I reply, tired and ready to leave. Tata Consultancy Services has offices all over India, I think to myself, but this has to be the worst. Simon and I are holed up in this uncomfortably small room, made even tinier by dark wooden walls. There are rows of identical brown and grey desks on each floor, with loose, lint-coated ceiling tiles hanging precariously above them. The air conditioners are creaky and the elevator that carried me here was so cramped that it gave me a brief attack of claustrophobia. I feel as though I've travelled thirty years into the past. I'm reminded of the way I felt when I arrived at the Indian Institute of Technology in south Delhi, expecting a state-of-the-art campus and instead finding grimy old lecture theatres.

Simon tries again. 'Gen X, Gen Y?'

'Yes, I know,' I murmur.

Generation X was the American baby-boom generation born between about 1965 and 1980. They were children who grew up with mass advertising and colour television, and who tended to be more liberal than their parents. Generation Y are the products of a later boom, born after 1976. Sometimes they're called the Internet generation. But I don't know why Simon would bring this up now. It's the end of the day, the offices will close soon, and I'm not taking in much information any more.

'In India, Gen Y is what I call the Why,' he says enthusiastically, going on to spell out what he means. 'W. H. Y. This is a generation that questions. That's why I call it Gen Why. The Gen Why category, my daughter's one.'

If I doubt that India's IT workers are creative and innovative, he tells me, then I need to think again. About the time that his daughter was born, he explains, India went through a major economic change. A quasi-socialist state since it became independent, in 1991 the country decided to open up to the world. Before then, it could take anything up to five years to get a telephone connection. Getting permission to travel abroad, even for one day, could take two weeks and repeated trips to the capital. But after 1991, government-owned companies, including the vast communications and steel industries, were privatised. Miles of bureaucratic red tape were ripped away.

American brands like Coca-Cola and Pepsi started appearing in Indian shops and a few years later, MTV launched its first Indian music channel. Millions of families entered the ranks of the middle classes, grew wealthier and began having smaller, nuclear families. So Simon's daughter, now at college, is coming of age in this new India, in which a fifth of people have a car or scooter, colour television, and a telephone at home.

But there is something more important happening in this generation, says Simon. 'The liberalisation of the economy has also developed into liberalisation of thought,' he argues. And this will turn India back into a nation of innovative geeks.

'We are now seeing the first generation of that liberalised economy,' he continues. 'And they are taught to think differently. They've got freedom of thought, freedom of behaviour, freedom of social etiquettes, freedom of profession. If I look at the last twenty-five years, the current generation is willing to think differently, experiment in thought, and the teachers are willing to accommodate experiment in thought.

'If I look back to my school days, some of the schools during my time, the teacher would say something and that was the end. There was nothing beyond. Today, teaching has taken a different paradigm.'

For these children, he says, sipping on lemon tea, life as a cyber coolie in an office cubicle is not enough. This is the generation that promises to change Indian technology companies from the backroom, maintenance outfits that they are now into innovative powerhouses like Microsoft and Google.

It hasn't been an easy transformation, he admits. 'Initially it was difficult for me because I was a bit of a replica of my parents. Fifty years back, I never questioned my parents. Today a child will inform you what he's going to do or what he's not going to do.' He pauses and looks around, unsuccessfully, for a picture of his daughter. 'My daughter fortunately informs me that she said she wants to do electronics engineering and work in communication. Partly it was peer pressure, because everyone was doing engineering, I knew her propensity for mathematics was very good, her ability to think logically. But I told her do what you want. She said to me that if things work out well, she would like to do chip design. She's only doing first year in engineering and she already knows what she wants to do.

'And that's not just a reflection of her,' he says, beaming with pride. 'I've seen this in students up and down the country. The mind works like a parachute – once it's open it works beautifully.' Even Tata Consultancy Services is changing, he promises me.

But I'm exhausted, and having already spent so much time in the dusty, quiet classrooms of the Indian Institute of Technology in Delhi, sitting near the drab cubicles in this office block is not helping to lift my scepticism. I'm not sure any more whether India is genuinely turning into a geeky, scientific society, or whether it's just hype.

Simon answers this question with a story. It's the famous Hindu legend of *amrita*, the elixir of immortality. Thousands of years ago, he says, a king was facing defeat in battle by an evil nemesis. Desperate for help, he went to the powerful blue-skinned god Vishnu, the protector of the universe. The god told him that, by mixing and distilling the oceans, he might be able to isolate a tiny amount of *amrita*. But churning the oceans seemed an impossible job, so the king went for help to the king of the serpents. Together they wrapped the serpent's enormous tail around a mountain in the middle of the ocean and spun the waters, finally extracting a tiny amount of the magical elixir. The king drank it and defeated his enemies.

'Now that ocean was really large,' says Simon. 'But India's engineering students are like the oceans. If you take just one per cent of that accumulated force, of all those hundreds of thousands of students, you will see that we can make this go forward.' This element of scale – unimaginable until this century – is what he believes will transform India's fate. This is a country of more than a billion, all passionate about brainy things, and now with a young, inquisitive generation coming of age.

'When I see 2020, which is ten years from now,' Simon adds, just before I switch off my voice recorder, 'you might find the next best application formed in any platform in any technology having an Indian connection, or an Indian origin. You might see an Indian invention.'

If that invention is going to materialise, there's one place it's likely to happen first. The journey to get deeper inside India's geeky core takes me 1,700 kilometres to the south, to the smaller city of Bangalore, now called Bengaluru.

This used to be a quiet, leafy place where pensioners migrated to retire, but in the last couple of decades, the sleepy city has turned into the technopolis of the subcontinent. Nobody even knows exactly how it happened, but its tree-lined boulevards are now choked with traffic and the city's parks have all but disappeared behind office blocks and shopping malls. A third of all of India's IT workers are here, not to mention almost all of the country's leading aerospace scientists, defence researchers, chemists and biotechnologists. The city used to be known as the next Silicon Valley, but it's already surpassed those ambitions and become a world-famous technology hub in its own right. It's stuffed with eccentric geniuses and freshly minted millionaires.

At least, that's until it comes to innovation. On this front it's scarcely better than a retirement village. To date the city hasn't produced a single company to rival Microsoft, IBM, Apple or Google, which make cutting-edge products and software that the whole world is using. Lucrative though it is, most of the work being done in Bengaluru tends to be day-to-day maintenance and routine software development.

To figure out if this might change, I arrange to meet Narayana Murthy, the 63-year-old billionaire founder of the software company Infosys. He is as famous here as Bill Gates is in the US and the appointment isn't easy to get. After weeks of waiting, I'm finally invited inside the inner sanctum of the company's campus. I'm hoping that, if anyone can help me understand how Bengaluru present squares with Thomas Simon's vision of India's future, Murthy can.

'Where?' the young taxi driver asks me. I'm standing on the roadside in Bengaluru's main shopping district, facing a flyover on which the city is building its first metro train line.

'Electronics City,' I tell him. 'I'm going to Infosys.'

He's impressed.

'I'm meeting Narayana Murthy.'

He nearly falls out of his seat.

Bengaluru feels like a different country to the rest of India. This street is lined with chic bars, including a NASA-themed one decked with blue neon lights and table legs that look like rockets. While it is almost impossible to find beef in most parts of India, because of the Hindu aversion to killing cows, it's blasphemously easy to get a good steak here. There is a French patisserie, American fast-food joints and Italian fashion stores. This melting pot of cultures and ideas has made Bengaluru an attractive place for scientific entrepreneurs. And Murthy was among the first of this young geeky cohort.

There are two satellite districts outside Bengaluru, each stuffed with multinational technology companies. The International Tech Park, on the east of the city, is a collection of gleaming glass towers with names like Discoverer, Navigator and Inventor. The other, Electronics City, is about twenty minutes further to the south. The people who designed it back in 1978 dreamed that it might one day become India's own Silicon Valley, and sure enough, every month the newspapers carry headlines about another business opening an office somewhere in this geeky galaxy. Hundreds are already here: American firms like Microsoft and IBM, the Chinese telecoms company Huawei, the European electronics firm Siemens, and of course each of the top three Indian IT companies, Tata Consultancy Services, Wipro and Infosys.

Murthy lobbied the government for fifteen years to have an elevated highway built between the centre of Bengaluru and his 82-acre company headquarters in Electronics City, to make the journey to work quicker for his Infoscions (this is what he affectionately calls his workers, as if they were scions of some great, dynastic family).

Luckily for me, the Hosur Road Elevated Expressway was finished a few weeks ago. And coasting down this smooth, sweeping road, hugged by curved silvery streetlights, doesn't just make my trip faster; it makes me feel as though I've arrived in Hollywood. It's comfortably high enough above the dirty mess of the garment factories and wooden shacks below to give the impression that I have travelled into a new world.

The Infosys campus is so iconic that tourists make this same journey just to look at it. From the expressway, three enormous glass structures emerge on the right like alien spaceships, the sun glinting off them. One is a pyramid modelled on the Louvre in Paris, made of diamond-shaped panels and surrounded by a moat. Another is a huge geometric glass shard, and the last is a cuboid with a circle cut out of it so you can see all the way through (locals call it 'the washing machine'). Here, they're as incongruous in their surroundings as the Kentucky Fried Chicken restaurant I saw once next to the Egyptian pyramids near Cairo.

Infosys is the first Indian company to build a megacampus like this one, trumping Americans firms like Google and Microsoft, which are famous for their employee perks, said to include expensive art on the walls, free gourmet food and sports clubs. The Infosys employee gym has a jacuzzi, snooker tables and dartboards. This campus also contains the largest television screen in Asia. The company spent a quarter of a million dollars on these buildings and has just finished another campus, in the neighbouring city of Mysore, which is even bigger and cost almost twice as much.

Murthy's assistant greets me at the gates and we hop into a buggy for a tour. More than 200 Infosys company buses take 21,000 employees to and from this campus every day. There are no cars inside the compound, only pink and blue bicycles, which are shared by everyone, and large golf buggies. It takes us a full fifteen minutes to do a circuit of the campus. It's as clean as the streets of Singapore, with not a single brick or a blade of grass out of place.

We pass an open-air basketball court and then a vast pavilion full of restaurants. 'We have all kinds of food. Indian, continental, Chinese – well, Indianised Chinese – Keralan, and lots of others' the assistant chatters on the way, our buggy skimming the glass pyramid. IT geeks are famous for their love of fast food: I spot branches of Domino's, the takeaway pizza restaurant, and Subway, the American sandwich chain.

We roll to a stop outside the building that contains Murthy's office. A simple, small, low brick tower, it's the plainest and oldest one on the campus. While we're walking up the stairs, I ask his assistant what she thinks of her boss. 'He's a perfectionist,' she says, flicking through an itinerary in her hand. 'When we organise an event, we don't just have a backup, we have a backup for the backup.' The look in her eyes tells me that he's something of a control freak.

'But he's brilliant. He's a brilliant man. What he has done is quite incredible,' she adds.

'So you say that being a geek is . . . good?' asks Narayana Murthy. He sounds unconvinced and I get the sense that the title of my book may even have offended him.

'It's a good thing. It's all about passion and inventiveness,' I say, reassuringly.

'Calling Indians geeks is like we have no other thing to do in the world except read books and all of that. You could be a little more generous.'

'So you wouldn't say that you're a geek?'

He thinks about it. 'Yeah, I suppose I am,' he laughs, pushing up his thick, dark-rimmed glasses.

Murthy is an enigma. He founded Infosys in 1981 with the equivalent of just $250 in his pocket, reportedly borrowed from his wife. By 1992, the company was making one and a half million dollars a year. And by 2008, it was worth more than four billion, making instant millionaires out of workers who had been given stock options. Soon the famous Indian comic series, *Amar Chitra Katha*, is to release his biography in pictures, which will make him a living comic-book hero. He's been called a geek god. But unlike so many of India's elites, he didn't inherit his wealth, and he has a personal philosophy that's famous for squaring business savvy with wholesome Indian values. He is one of the country's richest men but he subscribes to the same modest way of life as Mahatma Gandhi.

This lifestyle happens to suit his methodical and ruthlessly clutter-free system of organisation. When I enter his office, it's so stark that I think I've accidentally opened the door to a storage closet. But there he is, smiling, unassuming and ready to meet me. He's wearing a blue shirt with his trousers belted high around his waist. There are plain white floor tiles and high wooden bookshelves packed with hundreds of titles, each carrying a numbered sticker like in a library. In one corner is a meticulously tidy desk.

'I think simple living is good, because that's how you bring about a balance between greed and decency,' he says. 'If you look at the financial crisis that affected particularly the US and the UK, that's because of greed, right? What Gandhi said was, look, it is the inner force that has to bring a bit of balance between what is right and what is not right. Gandhi was one of the finest leaders India has ever produced and I think clearly one of the finest men.'

There is something endearingly eccentric about a billionaire who finds so much appeal in the simple life. To some extent though, like Murthy, the whole of India still lives in Gandhi's shadow. Most politicians here, for example, wear traditional homespun cotton fabric like Gandhi used to, instead of Western suits, to convince people that they are plain-living, ordinary folk like their voters. But Murthy is a man shaped as much by his experiences as by society's expectations. He doesn't just live by his ideals, he really believes in them.

According to Murthy, a single event turned him into the man he is today.

When he was young, he says, he was a committed socialist. And then, one holiday shattered his faith in the political left and set him instead on a career in the technology industry. It began when he was backpacking through Europe in his twenties. It was 1974 and he was unsure what to do with his life. 'They were the golden years to be in, you know. We were all lefties then,' he says nostalgically. He whiled away happy evenings chatting with French socialists in the same Paris café that used to be frequented by Jean-Paul Sartre, he hitchhiked through Rome and Israel, slept on railway platforms to save money and partied for free in youth hostels.

But these golden times ended abruptly. His face falls as he tells me the next part of the story. Murthy was 27 years old and travelling through eastern Europe on the last leg of his backpacking journey when he was stopped by police officers in a town bordering Bulgaria and the former Yugoslavia. It wasn't his fault, he tells me, it was down to a young man he met on his way, who thought he was criticising Bulgaria's communist government and reported him to the police. They imprisoned him without food.

Luckily for Murthy, at that time India had friendly ties with both the Soviet Union and the whole of the Communist Bloc. So the police set him free after five days, putting him on a train home. But the traumatic episode ended his love affair with socialism. On the long, cold train journey out of Bulgaria, he blamed the political left for the way he was treated in that lonely prison cell and he began to see socialism as barbaric. 'I said to myself that if this society treats friends like this, then I don't want to be part of such a society. That in some sense crystallised my thoughts and decided the issue for me,' he says. In the next few years he inched away from his old political ideals and became a technology entrepreneur.

But while he became a technology mogul, he never really abandoned his old life entirely. He seems to have married the spirit of his youth with the jetsetting lifestyle of a billionaire. His middle way, which lies in the grey area between asceticism and wealth, socialism and capitalism, Mahatma Gandhi and Bill Gates, explains the Infosys motto: 'Powered by intellect, driven by values.' It's a unique ideal that millions of Indian techies have come to admire.

'Do you think this philosophy is what made you a success?' I ask him.

'Well, I personally feel we have to be humble,' he says, adding that this applies to India as a whole. 'In fact I was with an Australian group just now and they asked me what is the secret to what you people have achieved? So I said that it is humility and five other attributes.' Murthy tends to list and number his thoughts. He counts each attribute off on his fingers, one by one. 'One is openness to new ideas, second is meritocracy, third is innovation, fourth is speed and finally, excellence in execution. We focus on these five attributes, with humility. The day we have hubris, well . . .' He tails off, tutting in disapproval and shaking his head.

'Really?' I ask.

'Yes, these five things.'

While these attributes may certainly apply to Infosys, however, I'm not convinced that they apply to the whole of the Indian IT industry. In the international arena at least, technological innovation has never been one of India's strong suits. Infosys became a billion-dollar firm in pretty much the same way as Tata Consultancy Services did; it capitalised on India's cheap and educated labour force to do workaday maintenance for Western customers. Even now, thousands of software engineers are stuck in dronelike jobs, and this has made the industry slow at inventing new products.

In fact, Infosys has hardly made a mark in the international science community. Compared to Microsoft, it spends a tiny fraction of its income on research and development. American technology giants often earn hundreds of patents on their inventions every year, but in 2009 the top three Indian companies between them only managed to file 150 patent applications. And fewer than a dozen of them were granted.

Secondly, while Infosys is known to be an egalitarian workplace, the rest of India is not. Much of society is split by both class and a Hindu caste system, which together often decide a person's fate before they're even born. A 2007 survey by the Bengaluru-based social scientist, Carol Upadhya, found that almost half of a sample of software engineers in Bengaluru came from the highest Hindu caste, the Brahmins, even though Brahmins comprise less than five per cent of the Indian population. And 84 per cent of the fathers of all the engineers in her survey had middle-class jobs.

So even though Murthy seems to live up to his rags-to-riches, humble story, I'm not persuaded by his vision for India's success. Outside his office, even the Infosys campus appears to be a far cry from his simple aesthetic. I get the feeling that he may be a man stuck between his ideals; between the impoverished, socialist India that he grew up in and the capitalist, technology-driven country that he is living in today.

It's a struggle that faces India as a whole.

'We are children of two cultures,' he says, opening a wider window into his mind. 'We work in the third world and we try and satisfy the most demanding of the first-world customers. We live in and we pass through the wretched world of Bengaluru, and then we pass through the autobahns of Germany.'

After his life-changing backpacking trip around Europe, about thirty years ago, Murthy tells me, he considered entering Indian politics. He wanted to change the country from its grassroots and help close the development gap between the global East and the West. In fact when India's former president came to the end of his term a few years ago, Murthy's name was even thrown into the ring as a potential successor.

'So why didn't it happen?' I ask.

'I thought I would join politics,' he says. 'But the rules of the game are such that people like me would not enjoy playing it.'

'Why not?'

'See, there's a lot of corruption. Second, fighting elections in India is not easy. It's based on caste, it's based on religion, it is based on who you are related to. When merit is relegated, when there is not a transparent system, when everyone who plays in the game competes and plays by those rules, this is what happens,' he says, bitterly.

'Whatever areas have been fudged by the government, we have not made as good a progress as we could have. Primary education, primary healthcare, nutrition, these are all areas that come under the government,' he says. Outside the Infosys security gates, past the manicured shrubs and at the end of the elevated expressway, literacy rates drop dramatically. For every thousand children born in India, forty-nine of them die as infants, compared to just six in the US. According to the Association for Democratic Reforms, a New Delhi watchdog that monitors the government, a third of all the members of the lower house of parliament have criminal charges pending against them, including charges of rape and murder. So scientists, engineers and entrepreneurs like Murthy have been forced to work around this crumbling infrastructure, building world-class businesses and training world-class workers despite India's endemic weaknesses.

The rest of India, though, seems a world away from the polished and well-managed Infosys campus. In the technology parks of Bengaluru, women and men like Murthy have been able to create idealistic communities in their own image – oases of technology, where ideas count for more than money and young geeks can live by their own rules. He may not have become a politician, but Murthy is the master of this place, with arguably as much influence as any member of parliament.

By the end of our encounter, I have realised why Murthy has supplanted grim reality with his imagination. But his grand dreams are more than a fantasy; they are visions that are shaping the future.

'I realised that if our youngsters have to go to our customers and deal with their problems confidently, they have to experience a quality of office environment which is as good as our customers. They have to have the best of infrastructure, the best of facilities,' he explains. Murthy has cherrypicked his favourite European architecture and recreated it in his own campus, so that his Infoscions can feel equal to computer engineers anywhere else in the world; he has lifted American chain restaurants and popped them into his food courts so they'll be familiar with Western tastes; he has imported the meritocratic atmosphere of Silicon Valley and tried to instil it in his own company; and he has brought his Gandhian ethics to the company, hoping to temper all of this with a sense of Indian simplicity and humility. And the result is that, unlike a decade ago, when India's software engineers were squeezed like sweatshop workers into cubicles in the gutter of the global technology industry, he has given them some hope. Now they're looking up at the stars.

The glass pyramid and the 'washing machine' outside his unadorned office are more than just expensive playpens for his Infoscions; they're almost a living manifestation of Murthy's geeky aspirations.

'In some sense they are temples of modern India,' he says, staring at his half-full teacup for a moment, dropping a white sachet of Infosys-branded sugar into it. 'They are temples that have raised the aspirations of our people. They are icons that have brought a new sense of enthusiasm to our people.'

I learn later that Tata Consultancy Services is also halfway through constructing its own megacampus, at the end of an IT corridor around an hour outside the southern city of Chennai. Ten thousand people are already working there. There will be more industrial metal in the new 70-acre site than in the Eiffel Tower in Paris, I'm told. The central spine is 320 metres long, stretched over a curving indoor water canal. It looks like a cross between the Sydney Opera House, a giant bumblebee and a south Indian temple. The workers describe it as the 'new Taj Mahal' and an 'icon building'.

'That's what temples are all about, right?' says Murthy. 'In the morning, I pray, and what do I pray? I pray that things are better for everybody in the world, for me, my family, everybody. And when I do it, I get a sense of satisfaction, a sense of confidence.

'We lost our tradition in science, mathematics, astronomy, for a thousand years, but I have confidence that we will reclaim it.'

Dr Manish Gupta has a new invention.

Wearing a baggy red sweater and the shadow of a moustache, he's the director of IBM's laboratories in India. The giant cylindrical offices he works in, filled with padded grey cubicles resting on pastel carpets, are a treasure chest of the gadgets that you and I may not know about now, but may be using in a decade. Headquartered in New York, IBM is one of the world's largest technology companies, spending six billion dollars a year on research and development and regularly topping the list of most patents awarded in the US. In 2009, it got almost 5,000, beating its competition by miles.

Here, ideas are like gold dust and engineers are like magicians.

I try to take a peek behind temptingly closed doors on the way to Gupta's room. In 1953, IBM designed the first popular general-purpose computer. In 1981, it made a microscope that allowed scientists to see and move individual atoms for the first time. And it was one of the first companies to develop devices that could take human speech and turn it into text. Its researchers have won five Nobel prizes and eleven more have made it into the US National Inventors Hall of Fame. The reason I'm here is that the company's India research laboratories, founded more than a decade ago, are now making some of IBM's most important inventions.

Most of the people working at IBM have PhDs, I'm told, and if they don't, they're encouraged to get one. This place is a scientist's dream. Unlike Infosys and Tata Consultancy Services, IBM pumps so much cash into research that this is one of the few places in India where they can spread their creative wings and experiment with radically new ideas.

Manish Gupta is one of dozens of computer scientists to do just that.

Like Narayana Murthy and Ananth Krishnan, Gupta is also an Indian Institute of Technology graduate, from 1987. But he has the hint of an American accent and the habit of saying 'right?' at the end of his sentences. He spent fourteen years at the IBM TJ Watson Research Center in New York, he tells me, heading up research on software for the IBM Blue Gene supercomputer, then the fastest supercomputer in the world (it could make just over 36 trillion calculations per second).

Gupta wasn't the first Indian to cut his scientific teeth in the US. Entire generations of IIT graduates ended up there, the country becoming almost a training ground for India's most innovative scientists and engineers (it was known in the 1980s and 1990s as the 'brain drain'). In fact Indian geeks, nerds and dorks have become a regular feature of American popular culture. The character of the gawky, bespectacled computer scientist in the 1986 cult-classic Hollywood robot movie *Short Circuit*, was an Indian. And even though Apu Nahasapeemapetilon in *The Simpsons* is a hardworking grocery store owner, he also has a PhD in computer science.

More seriously, Indian immigrants and Americans of Indian descent are now a major force in Silicon Valley. The founder of the software company Sun Microsystems, for example, is Vinod Khosla, an Indian-American who studied at the IIT in Delhi. Sabeer Bhatia, co-founder of the popular email service Hotmail, which was bought by Microsoft on his twenty-ninth birthday for about 400 million dollars, is also an Indian-American. And the co-inventor of the Universal Serial Bus (popularly known as the USB), which allows hard drives, mice and other gadgets to be plugged into computers, is an Indian-American working for Intel in California. In fact, at the turn of the millennium, there were 20,000 Indian immigrants working as scientists and engineers in Silicon Valley.

For India, the loss of all this brainpower may have held the country back once. But the drain is now getting filled back up, thanks to people like Manish Gupta choosing to return. In the second half of 2008, Vivek Wadhwa, an adjunct professor of engineering at Duke University in the US, polled 1,203 high-flying Asian workers in Silicon Valley who were planning to leave about their reasons for de-emigrating. More than two thirds of the Indians said it was because professional opportunities were better back home.

'What excites me,' says Rajdeep Sahrawat from the National Association of Software Companies, 'is that the Indians who left India in the 70s and 80s, and we all cried about it because we were losing talent, now they are the guys who have reached a level of fairly senior-level responsibility in the West, especially in the US. They are the heads of computer-science departments. Now they're the ones coming back to India and saying, can we do something? I think these are the guys who will have the diaspora effect, a reverse brain drain.'

More surprisingly, the multinationals that they worked for in the US have also followed them here. 'There is a fairly large proportion of Fortune 500 companies which have set up research and development centres in India,' Gupta says, turning in his swivel chair to face me. In fact, there are 800 such research centres here now. 'If you look at the top IT companies, if you look at the likes of IBM, Microsoft, Google, Yahoo, we all have research centres in India, right? Initially research and development centres of this kind were places where researchers would essentially help augment things being done at the corporate headquarters, but increasingly you'll find examples of innovations that have come out of the research labs here. I think we've always been innovative but the opportunity for innovations now is becoming a lot more obvious.'

Their investments are already paying off. Intel, the company that makes the microprocessors that are inside most of the world's computers, for example, opened a design centre in Bengaluru back in 1988. Within a decade, the workers there helped build a new computer chip. A few years later they were part of a team that created a prototype of a super-small, energy-efficient processor, which uses only as much power as two ordinary light bulbs, two thirds less power than other chips use.

Gupta believes that returning Indian brains and multinationals like IBM are breathing new life into the Indian technology industry. 'One of the reasons why you have research labs in different geographies is because of the unique needs that people see in those areas,' he says. There are some problems that are unique to India and China, so the innovation has to happen first right here. What we're trying to do is apply technology to help solve some of the day-to-day problems of these people.'

Gupta finally reveals his new invention.

'We call it the Spoken Web,' he says enthusiastically, waving his mobile phone in front of my face. 'Let me explain. See there's absolutely no doubt that mobile phones have caught on and are continuing to catch on at a much higher rate. Not just in India, if you look at Africa, countries in Latin America, there are enough places in the world where there is a significant rate of mobile phone adoption, more than Internet and computers. So we think this is an extremely powerful technology.'

He shows me the scientific papers that explain how the Spoken Web works. The basic idea behind it is that, in the same way as people use written words to surf the web on their computers, a voice-based web lets them use spoken words to surf spoken information on their mobile phones. Pieces of speech are linked to each other in exactly the same way as pieces of text are linked to each other in the World Wide Web.

It has turned into one of IBM's biggest projects, receiving a hundred million dollars of funding over the next five years. But unlike IBM's other inventions, it was born out of a problem specific to Asia and Africa. In India, computers are so rare that only about 52 million people in India's urban areas use the web regularly. Many more, however, do own simple mobile phones (the average Indian earns under 4,000 rupees a month, putting the cheapest mobile handset, which is about 800 rupees, or a recycled handset, just within their grasp). The telecommunications revolution that began after India launched its first satellites, teamed with mobile phones, has since transformed the way India works. Mobile communications have leapfrogged land-based ones. There are already more than 400 million mobile subscribers and estimates suggest that there could be as many as a billion by 2014. Call prices have fallen so low that a minute on the line can cost as little as half a US cent. So in this environment, where computers may never proliferate, the Spoken Web could boost the number of Internet surfers by hundreds of millions.

Gupta is almost romantic about the idea. 'I mean our whole civilisation is kind of almost founded on this concept of the spoken word. Our scriptures were not written, they were passed down by mouth.'

In truth, though, the Spoken Web is not just designed for people who prefer not to read text or can't afford to, but for those who don't have a choice. India is home to 350 million illiterate adults. And they speak a mind-boggling number of languages, many of which are unrepresented on the Internet. The last Indian census recognised more than 1,500 different mother-tongue languages, some so different from each other that a Punjabi speaker from north India, for example, would find it impossible to understand a Tamil speaker from the south. With the Spoken Web, even a dialect that's only spoken in a single village can have its own network. On top of this, there are also 15 million blind people in India who would benefit from the invention.

The hope is to finally bridge the digital divide between the rich and poor by bringing something similar to the Internet to everyone who has a telephone and can talk. Gupta calls it a 'parallel web for the masses'.

He explains how it works. Academic research into voice-based Internet started back in 1991, when a researcher at the Massachusetts Institute of Technology, Barry Arons, invented the word 'hyperspeech'. The text-based Internet uses hypertext, in which words and pictures are hyperlinked to other websites by encoding them with the location of a new and separate piece of text or another picture, using lines of computer code. Hyperspeech would work similarly, Arons wrote, except the hyperlinks would exist between audio items instead.

He imagined a network made of speech files in a database, which a person could navigate using spoken words. The computer would recognise what they were saying using electronic speech recognition and then synthesise helpful phrases to guide the person to the next file. Sets of information could be grouped by topic, making it easier to find relevant things.

It sounded like a simple idea, but it posed an enormous technical challenge. In 1991, neither the hardware nor the software existed to make it work (mobile phones were hugely expensive and the World Wide Web had not yet taken off). Arons prophetically suggested, 'Speech interfaces are particularly appealing in the context of small portable or handheld computers without keyboards or large displays.' It would take almost two decades for technology to catch up with his radical vision.

Now mobile phones with smart chips and graphic displays do exist, and there are also easy ways of collecting and digitising huge amounts of audio data. In India, they have taken these two technological trends and turned Arons' vision into a reality.

A few years ago, a team of five researchers at the IBM India Research Laboratory invented a set of rules known as the Hyperspeech Transfer Protocol, or HSTP. This is similar to the Hypertext Transfer Protocol, or HTTP, which is what website designers use to organise words into hypertext, with hyperlinks to other sites. With HSTP, the Indian researchers solved the difficult technical problem of how to link one piece of audio to another, by using telephone numbers like website addresses. The phone network is essentially a vast system of switches, which connect each person who has a telephone to every other person who has one. A spoken website has its own phone number, which you can call. And moving through this Spoken Web is like clicking one switch after another in this vast network, being transferred from one number to the next.

Since mobile-phone users don't have a computer mouse to click with, the team also came up with four commands they could use to move through this audio network: next, previous, browse forward and browse back.

For the first seven months of 2009, workers at IBM and researchers from Berkeley and Stanford University tested out a basic Spoken Web system on fifty-one people in Gujarat, a state in western India. Their aim was to find out whether a typical Indian farmer would use it. Their human guinea pigs were young, had small-scale plots of land, basic levels of education and had barely or never used the Internet. This spoken network included a question-and-answer forum and an announcements board for messages about agriculture, animal husbandry, weather and food prices. But of course, as any Internet surfer knows, the web is as much for entertainment as for useful data. So the team also added the entire archive of a popular local radio show, broadcast in Gujarati.

Access to the sites followed a simple format. A farmer would dial a number and be given a set of options, which he would select on his mobile phone keypad. This would take him to the next spoken site, and so on. It worked in almost the same way as an automated helpline, which banks and utility companies often use to process customer calls. It's known in the industry as interactive voice response.

The pilot was a success. There were 6,975 calls to the Spoken Web in seven months, and farmers posted 610 questions. The experiment also threw up some fascinating results, challenging preconceptions about how poor, illiterate people approach new technologies. The expectation was that there would be uncertainty and hesitancy about using the system, but in fact it became popular straight away.

People who surf the Internet on their computers follow a certain pattern – usually there will be a few highly active callers, others who dip in occasionally, and a handful that don't use it at all – and the users of the Spoken Web in Gujarat did exactly the same. One farmer recorded a question about a hot wind that was wrecking his millet harvest, to find it helpfully answered by another millet farmer, who suggested that he plant a different variety on the edges of the crop to shelter it. A family left the sounds of their newborn child so their relatives could hear it. The callers even moderated each other's comments, just as in an online forum. In fact, when one man posted a marriage advert on the message boards, other villagers exposed him as an untrustworthy prospect.

'Those farmers were not even aware, I mean they didn't even

know what underlying technology was being used in the system. But they took to it. In some cases, their farming output has gone up, in some cases the use of pesticides has gone down. We're really proud of doing this work,' says Gupta.

While the pilot project in Gujarat was small, with just a handful of links, in the long run IBM envisages a Spoken Web as unlimited as the written web. Instead of a few options, there might be links between hundreds of businesses and thousands of people. A caller could buy a shirt on one site, be linked to a payment site to enter their bank details, and then be connected to a message board to tell everyone about it. When the invention has matured, IBM scientists hope that businesses will develop their own audio websites and friends will share gossip on their own spoken networking sites.

There was one problem, though, which Barry Arons also flagged up back in 1991. It's impossible to browse through sound in the same way as browsing words and pictures. 'When you look at a webpage you can immediately pick things out, and so on, whereas in audio there's an inherent sequentiality to it,' explains Gupta. Every audio file needs to be heard one at a time. Organising the Spoken Web without getting lost in a tangled maze of words and music might mean having to transcribe all the speech and then lay it down on paper.

These are obvious wrinkles that still have to be ironed out, but even so, Gupta is convinced that the Spoken Web will be India's next information frontier.

An IBM employee takes me to the coffee lounge for a break.

After everything I've seen so far, it's a relief for me to finally find some examples of genuine creativity and innovation among Indian engineers. But what bothers me is the *kind* of technology they've developed here. If the Spoken Web is meant to bring the Internet to the illiterate masses then, in effect, this is a technology that is bypassing the need to read at all. What does that mean for the country's development?

I call Rajdeep Sahrawat at the National Association of Software Service Companies for his views. What troubles me, I tell him, is that the Spoken Web, even though it is a promising homegrown technology, seems to be plugging a gap created by the government's failure to provide basic education to more than a third of India's population. 'Wouldn't it be better to just educate everyone?' I ask. 'Don't you find it strange that technology companies are making such a high-tech solution to this problem? I mean, if people can't read, why go to all this trouble when you could just make them literate?'

'I wish it was that easy,' he replies. Sahrawat is a straight talker and more pragmatic than I am. 'People of my generation are very realistic. I know that the government is doing a lot of good stuff, but I also know that the government is not going to solve all the problems.'

'Can throwing expensive technology at a problem really fix it?'

'You know there are 300 million people in India who don't have a bank account? Most of these people are illiterate and they don't have an address. They can't even leave a fingerprint because they work in the fields and their fingerprints have worn off. How are we going to solve that problem? The conventional paradigm is low tech for the poor at a low price. But we have to leapfrog that. I think some solutions in India will be low-tech but some will be very, very high-tech.

'There's this prism of thinking in the West,' Sahrawat continues. 'There's this quintessential question which comes up, which is when will India create Windows? When will India create Googles? I think that's the wrong question. Creating a Windows is not our problem. Creating a Google is not our problem. In India we have other problems to solve. Our problem is our scale. Our problem is feeding and educating a billion people. You need to go to the rural areas, and see what it is like.'

I think about this as I leave IBM. The idea that a greater

number of Indians may eventually own mobile phones than can write their own names still seems to me dystopian, but in the absence of a strong government, and with millions of qualified scientists and engineers looking for problems to solve, perhaps this is what progress looks like. I can't expect this twenty-first century geek nation to look the same as the scientific superpowers that went before it.

The buzz surrounding inventions like IBM's Spoken Web has become so feverish that India is gradually becoming a hub for voice-based technology. I take another taxi ride, this time through Bengaluru's residential streets, to meet the founders of a tiny startup who are creating waves in the global technology community for their own contribution to this emerging field.

Both in their mid-thirties, Yusuf Motiwala is an electronic engineer wearing a shiny black shirt and jeans, and Apul Nahata is a computer scientist. I find them in a two-room office on top of someone's house, opposite a mosque and a small newsagent. Motiwala got his masters degree from the Indian Institute of Technology in Mumbai, he tells me, before moving to the US to work for Texas Instruments, an American company that makes the computer chips that go inside mobile phones.

They're like mad inventors, I think when I climb the narrow steps to their bare office. Motiwala's laptop is squealing like a banshee.

'Something wrong?' I ask.

'It's strange, no? It does this sometimes' he smiles, nervously, before shouting into the room next door. 'Apul, can I borrow your laptop for a moment? Mine is screwed up.' He presses the computer between his palms and gives it a hard thud.

'Does that work?' I say sceptically.

'Yeah, it usually does.' Their company, TringMe, he tells me between laptop thuds, is only two years old so they haven't moved to fancy offices yet. Judging by the bare walls (save a whiteboard scrawled with numbers), it looks like they could use some help.

But then, appearances are deceptive. Motiwala and Nahata made more than half a million dollars last year and they have at least a million users worldwide, most of them in the US, Europe and the Gulf.

'We're the first company in the world that allows calling directly from your browser,' he continues, now turning his laptop upside down. The squealing noise stops. Motiwala asks for my mobile phone number and types it into the computer keyboard. A second later, my phone is ringing. His laptop is calling me. That's not all, he says. His laptop can ring any website that uses sound and communicate with it as quickly and simply as making a phone call.

In the most recent national elections, Indian politicians used TringMe to make 10,000 simultaneous calls to potential voters. All they had to do was type the phone numbers into the website, leave a prerecorded message and it took care of the rest. The government health department also uses prerecorded, scheduled phone messages to let HIV patients know when to take their medication.

Using the Internet to make phone calls is nothing new (applications like Skype and Google Talk have being doing it for years) but the problem to date has been that designing voice-based software is not easy. It's a lot simpler for designers to make a text-based website with a few lines of code than to spend hours trying to program a computer to use audio instead. Motiwala flashes a picture of Einstein on his laptop (he's doing some show and tell for my benefit). He jokes that you 'had to be Einstein' to program a voice website before TringMe came along.

'What we came up with is, why not make voice programming as simple as going into a simple program and printing "Hello" and it speaks "Hello"? Right? This way you don't have to be an expert in programming voice, you just use whatever language you're comfortable with and we take care of the voice conversion.' The easier it is for coders to design voice-based applications, the more they will be tempted to do it and the more popular voice-based web will become. 'Ours is a much simpler package, a much more powerful package, which any developer can understand. Because it's so simple, look at the areas across which it has been applied. Anyone can create a web application and it will just speak. People really liked this idea, the whole world liked it, because nobody ever thought that voice programming could be that simple,' says Motiwala.

The way TringMe works is actually quite simple. Web developers tend to use only a handful of popular programming languages to build new sites, including HyperText Markup Language (HTML) and Hypertext PreProcessor (PHP). So Motiwala and Nahata decided that, rather than creating an entirely new language, they would invent a pared-down, parallel version of PHP, which they call Voice PHP. The beauty of this system, they tell me, is that it takes just three lines of computer code to make a conference call between three different phone numbers plus a website. Without Voice PHP, this task would usually take dozens of lines of code.

'We always wanted to start something new. This was a dream, an old dream. It was a pretty powerful idea and we had enough experience to turn this idea into reality,' says Motiwala.

Their idea could be one of the missing links that helps turn the Spoken Web from a laboratory experiment into a mainstream technology. He reels off a string of names already using the system. 'There's Facebook,' he says. 'And IBM uses VoicePHP in its call centres. Even Infosys is using us to power their phone banking system.'

74

Motiwala and Nahata remind me of geeky Silicon Valley entrepreneurs, working out of shoebox-sized offices and risking their life savings on their ideas. On a shaky shelf at the back of their office, next to a small Hindu shrine, is an award that they received a few days ago from the National Association of Software Service Companies for being India's most innovative startup this year.

I ask the pair why they chose to start their company in India, rather than in the US, where so many of their customers live. Cost was one factor, they admit. India is far cheaper. But Bengaluru has other things going for it, says Nahata. Dozens of technology startups like theirs have popped up in the city over the last few years. Young engineers and designers regularly meet up to share ideas. 'It's a very good city. There's a good ecosystem here,' says Motiwala.

Indian researchers and engineers are becoming more creative, adds Nahata. 'People say that Silicon Valley is where it's really happening, but Bengaluru is really happening too. It's happening.'

There's a joke that sums up how the Indian Institute of Technology in Delhi works. And in typical nerdy fashion, it has a technical theory behind it. Students work on the rocket principle, they say. They only get going when their asses are on fire.

I'm returning to the institute for a couple more days in an attempt to dig out final proof of the existence of the innovative, geeky generation that Thomas Simon from Tata Consultancy Services told me about. Most students are revising as usual, so I end up in an electronics laboratory, talking to Nishant Ranka, the electrical engineeering senior who I shared a snack with the last time I was here. 'Visitors usually aren't supposed to be up here,' he whispers. 'But they won't notice.'

The laboratory, in one of the college's newer buildings, is different from the lecture rooms and offices I saw before. The entire block has elevators and air conditioning. The big, sunlit rooms are lined with rows of workbenches, messy with coloured wires, soldering irons and oscilloscopes. In one corner is a bicycle hooked up to an electric motor.

'The thing is, it's not that students aren't innovative,' Ranka says, perching on a table, grabbing a half-finished circuit board and pulling a loose wire on it. 'We just don't get enough free time. When I talk with my friends, they have such bright ideas, but then don't execute them. If you don't help them overcome their inertia, students just aren't interested.'

A few of his friends, who were glued to computer monitors when we came in, join us. 'Things are changing,' one says. In fact, I learn, Ranka himself is part of a group of students who are stirring up a revolution.

Less than a year ago, he tells me, he and a handful of engineers got together to set up a new society designed to give everyone the chance to build practical skills in their own time. They were tired of the theoretical, nose-in-a-book culture developing at the IIT and they wanted to stir things up. So they created a free space for tinkering with gadgets and learning new computing languages without the pressure of tests. The first club of its kind on campus, it was named Technocracy.

They hadn't expected to get much interest at first. But the IIT community was so excited by the idea that they signed up 300 students in the first two days. Inspired by the desperate desire to get their geeky juices flowing, the group lit a fire under the ass of the entire college. They now have their own website. 'Technocracy gives you opportunities to learn and enhance your skills, utilise them to do fantastic work in your projects or to build your own dream, and finally share your experiences with other bright minds like you,' says the homepage. 'Keep up your enthusiasm and the world will be yours.'

Ranka's eyes light up when he talks about it. 'In Technocracy we have six teams, working on things like MATLAB [a high-level computing language for engineers], microcontrollers, robotics and web development,' he says. 'The main purpose of the workshops is that people get their hands moving. Labs here close at five pm, but our technicians are quite talented, so we asked them to stay late to help us. On the third workshop we held, people were given a microcontroller to build. We showed them how to solder, how to put the circuit together, and they enjoyed it so much that one of these workshops went on for more than five hours.'

It's an underground movement that's gathering momentum. Of the sixty-one patents granted and 205 pending patent applications filed for inventions by students and staff at the IIT in Delhi in its fifty-year history, forty were made in just the last year. The rocket is taking off.

The more I investigate, the more stories I hear from across the country about young scientists and engineers who are stretching their creative muscles for the first time. Five students at the IIT in Kharagpur, West Bengal, have invented a biological battery, which uses microbes to clean untreated sewage while at the same time generating electricity. The IIT in Madras has set up an incubation unit for new inventions. Meanwhile at the IIT in Mumbai, the annual Techfest for student inventions has begun attracting as many as 60,000 visitors a year. The last festival had a workshop on three-dimensional animation and another one on artificial intelligence. Later this year, the International Robot World Cup is taking place in Bengaluru, pitching fifty-two teams from around the world in a contest between their best football-playing bots.

Even the IIT faculty, I find out later when I quiz the dean, has started dusting away the cobwebs. Firstly, they have reduced the number of teaching hours, so students are not so overworked and have more time to explore their own ideas. Millions of rupees have also been poured into new engineering projects and state-of-the-art laboratories. Research funding at IIT Delhi tripled between 2003 and 2008, which means that nowadays, while the windows may be dirty and most rooms have ceiling fans instead of air conditioners, few students complain about the lack of books or equipment.

'The quality and quantity of research output has definitely increased. Another important thing is the number of industries sponsoring projects and setting up labs here,' the dean tells me. 'Now we're introducing a pyramid structure so that more students stay beyond undergraduate level. For example, our undergraduates can now join the postgraduate programme early and get a PhD in less time.'

A few months ago one of India's biggest bakeries, Harvest Gold, reached out to the public to help them design better crates that wouldn't break when they were making bread deliveries. Ten teams from IIT Delhi sent in their plans. 'Redesigning the crate was a chance for me to apply the theoretical constructs that I learn,' says Manish Chauhan, a 24-year-old postgraduate mechanical-design student who entered the Harvest Gold challenge and came third. And he feels encouraged to keep experimenting. 'Things like this are inspiring us to think about society and share knowledge. It provided the platform, and now we're taking it forward. Some students have come up with the idea of opening an astronomy club. Others want to start a gizmo club, you know, for gadgets. The culture around here is shifting,' he says.

Manish is in a rush to get to the library to revise for exams like everyone else, but he keeps our conversation running for another few minutes. He wants to talk about his latest project. It's a novel that he's been writing in his spare time with the help of one of his professors.

'What's it about?' I ask him.

'It's like science fiction, about robots. I love robotics,' he says.

Maybe Thomas Simon was right after all, I think to myself. Between small nerdy startups like TringMe and the shoots of scientific exploration blooming here inside the IITs, the country does seem to be on the verge of a transformation. As he chats about his book, Manish reminds me of the lovable Indian roboticist in the movie *Short Circuit*. There's nothing dronelike about him. He's a real geek.

CHAPTER 3

THE LONG-LIFE BANANA

On my last day in Bengaluru I leave the technoparks and multistorey shopping malls behind so I can explore the older parts of town. I stop at Koshy's, a 70-year-old café on St Mark's Road with whirring ceiling fans, faded wooden columns and smartly-dressed waiters. In the past it's been a celebrity hotspot for thinkers and thespians (the owners claim that even Queen Elizabeth and Prime Minister Jawaharlal Nehru came here when they visited the city decades ago) but these days it's more of a bohemian hangout for students and journalists.

It's a good place to hear a story. And it doesn't take long before a guy on the neighbouring table interrupts my lime soda with one. 'You want to know where this name, Bengaluru, came from in the first place?' he asks, nodding at my notepad. 'See there was this king long ago, and he went out hunting. He got lost and hungry. But later he met a holy man, who appeared from nowhere, and this holy man gave him some beans, saving his life. And the name of those beans was Benda Kaluru. Benda Kaluru . . . Bengah-looroo.'

'So the city is named after *beans*?'

'Yes, boiled beans.'

I find it ironic that somewhere as cosmopolitan as Bengaluru could have its roots in a myth about some magic beans. But in this largely agricultural nation, it does feel as though there's something miraculous about the city's transformation from a leafy retirement town into a gilded technopolis. Although of course it didn't happen magically overnight, it almost feels as if it did.

And now the remainder of the country is restlessly waiting for its own beans. Nehru had dreamed that scientists and engineers might eventually bridge the divide between the rich and poor. Instead as cities like Bengaluru become wealthy on the back of the software industry, rural Indians are being left further behind. More than half a billion still make their living from farming, directly or indirectly, and most grow only enough to feed their families. Bad harvests often cause deaths.

So while the efforts of computer scientists like those at TCS, Infosys and IBM may help, what farmers really need is a revolution.

'Did you know that India is the biggest producer of the banana?' asks Dr Pravendra Nath, a molecular biologist, wearing a trendy pink shirt and moustache. I'm inside Nath's laboratory at the government-funded National Botanical Research Institute, where I've heard that groundbreaking work is happening into a new generation of super-crops. They promise to change the lives of farmers across India.

The atmosphere is young and relaxed, like a university common room. Young scientists are zipping around us, swirling liquids in test tubes and gently tending to rows of experimental banana plants.

The reason that they're working on bananas is that, in India, this is the one fruit that seems to be sold on almost every street and, in some areas, served with every meal (it's good for calming the belly after spicy food, which means Indians can't live without them). For banana growers though, like all growers, the big problem is that fruit and vegetables rot. Bananas can be found piled high on handcarts, in yellow bunches that turn blacker every day until they have to be dumped. 'The shelf life is very short,' says Nath. In fact as much as a third of all the fruit grown in India goes to waste because of it. 'So we started working on the banana, thinking that if we could reduce the process of spoilage, if we can stop that, then we can extend the shelf life.' If it works, the same technique could be applied to other fruits.

The way Nath's team of researchers is approaching the problem is groundbreaking. They're not working with chemical sprays, fertilisers or novel banana-storage containers. They're applying a technology known as genetic modification, or GM, which works by tinkering with the smallest biological building blocks of the banana: its DNA. This is the substance inside all living cells that acts as a kind of instruction manual, telling them how to behave and what to become. Their job, he explains, is to fundamentally restructure the banana so that even before the fruit begins to bloom, it's 'programmed' to spoil more slowly.

Nath points through the glass partition in his office to the benches next door, loaded with flasks and test tubes. 'The whole lab is involved in what we term as ethylene biotechnology,' he explains. 'Ethylene, it has several roles, but it is involved in the shelf life of the fruits, vegetables and the flowers.' It's a natural hormone found in tiny levels inside plants, controlling exactly when fruits ripen, flowers open and leaves wither and fall. When a banana starts to ripen, it pumps out ethylene gas, which in turn triggers the production of even more of the gas, in a loop until the fruit has softened and then completely rotted away.

Bananas are a good test fruit for scientists because, compared to others, they're particularly gassy. Natural bananas give off so much ethylene, for example, that an apple stored next to a bunch of bananas will ripen even faster than usual, as it sucks in its cloud of gas. It's the reason that bananas ruin fruit bowls.

'The idea is that if you can suppress the production of this ethylene, the spoiling process will be suppressed too,' Nath continues. The principle is not entirely new. For years, florists have dipped flower stems in a solution of powdered silver nitrate, which artificially keeps ethylene at bay, so the petals stay fresher for longer. But Nath's bananas don't need to be dipped in anything; his team is trying to subtly change the banana's DNA so that it makes less of the rotting gas in the first place.

In essence, Nath's team is rewriting the banana's instruction manual. It sounds simple enough, but the tiny scale of DNA (it's only two billionths of a metre wide) and the fact that genetic science is still in its early days, makes genetic modification one of the toughest branches of engineering.

Scientists didn't even know what DNA looked like until sixty years ago, when the Cambridge University researchers James Watson and Francis Crick, and the British experimentalist Rosalind Franklin, suggested that its structure might resemble a double helix, or in other words a twisting ladder. The tiny segments of DNA are known as genes and carry information about how the living world is built. Every person has around 23,000 genes, and plants have typically tens of thousands.

One of the most important things that researchers have learned since Watson, Crick and Franklin's time is that, as far as our genes are concerned, animals are not so radically different from plants. Since all life on earth evolved from the same basic organisms billions of years ago, we have a lot of genetic material in common. Half of the genes in a banana, for example, are also in human DNA. This is what allows genetic modification to work. At this level, we are all so similar that we can understand how genes work by comparing species, and sometimes genes for certain beneficial traits can then also be transplanted from one species into another.

In fact, Nath's team is only the latest in a long line of researchers to attempt this kind of engineering feat. The first GM crop appeared way back in 1992. It was a tomato, branded the Flavr Savr, which was developed and sold in the US. The Flavr Savr had a slightly longer shelf life than an ordinary tomato because of an extra gene that delayed the production of an enzyme that controlled how quickly it softened.

Unfortunately, the tomato was a miserable failure. Firstly, it wasn't particularly flavourful. Secondly, shoppers were too frightened to buy it because of its unusual scientific pedigree. The company that made the Flavr Savr also suffered so many hitches that, in the end, it couldn't continue production. It marked the birth of a powerful anti-GM lobby in the US and Europe, campaigning against these so-called 'Frankenstein foods'.

But in the years that have followed, the quality and reputation of GM crops in the US at least has improved dramatically. These days, herbicide-resistant soybeans and rapeseed, and insectproof corn, are so popular on American farms that they make up more than half the total growing area for each crop. And scientists like Nath at the National Botanical Research Institute believe they can produce similar super-crops for India, just as in the US.

His young team is tackling the shelf-life challenge slightly differently from the researchers who developed the original Flavr Savr tomato. Instead of introducing a foreign gene, they're trying to find the exact segments of DNA responsible for ethylene production so they can suppress them.

'During this whole ripening and softening process in the banana, more than 10,000 genes get activated. You have to pick the gene of your choice, the right gene that you will use. It's the gene that should not only extend shelf life but it should not compromise on the quality of the fruit or the quantity you get from a crop. That's a big job!' he says.

If it works, the possibilities are endless. A banana that stays fresh for just a few days longer than a normal one would mean thousands of Indian banana growers could stretch out their harvests, sell their crop for longer and be sure their stock was still good to eat. Bananas growing in south India could even be driven to the north of the country without rotting on the way. There is another argument about whether transporting food thousands of miles from where it's grown is a good idea for the environment, but it does mean that millions of bananas could be rescued from rubbish bins.

Even if the science is a success, however, getting these bananas onto farms is another matter. Even though Nath is tantalisingly close to a breakthrough, there is a growing group of activists in India who do not want the long-life bananas to be created at all.

Part of this resistance is down to the same fears that arose in the US a few years ago, when consumers became worried that GM crops like the Flavr Savr might harm them or damage other crops and natural flora. But this is a land dependent on agriculture, in which the politics of food is not about lifestyle choices but about subsistence livelihoods. Farmers desperately need fatter, more profitable harvests. So the reason why thousands of protestors here are still convinced that genetic modification is both unsafe and unnecessary goes deeper than concerns about the environment.

In a way, the controversy draws a line between India's traditional past and its modern, technological future. Away from this slick laboratory and its educated geeks, India is home to millions of poor farm workers whose lives haven't changed for a hundred years. They plough their fields with bullock carts, rely on the monsoons to water their seeds, hope for good weather, and pray they survive another season. To understand their opposition to the science behind the long-life banana, I pack my bags and decide to head to the small farming village of Kuljeri in the district of Vidarbha, in the western state of Maharashtra.

The flat landscape stretching to the horizon ahead of me is like the Arizona desert. The ground is coated with a dusty, light brown soil, which is scattered with rocks and the odd tree stump. It gets as hot as 48 degrees Celsius here in the summer, with only a few months of rain before November. Barely a thing survives without help. The only flashes of colour come from the red, sickle-shaped palash flowers, which manage to push their buds through wiry twigs and dot the dry fields like fat drops of blood.

'Small-scale agriculture in India is maximum employment generation work,' says Vijay Jawandhia, a local landowner and farmers' rights activist who has offered to escort me through the villages of Vidarbha. He flaps the tails of his pressed white kurta pajama to keep him cool. It's unbearably stuffy inside the car and he also has allergies. 'Sometimes this happens to me!' he says, sneezing violently and wiping his teary, bloodshot eyes with a handkerchief. Jawandhia used to be a city dweller and studied chemistry at university, but he moved back to his family farm when his father died about thirty years ago. He's a big man now, not only in stature but also in influence. Farmers around here look to him for advice on which crops to grow, new agricultural practices and legal counsel. And I'd never be able to navigate the network of bumpy, signpost-free roads and find Kuljeri without him.

After a few hours on the rocky road, we're getting close to Kuljeri. Soon I start to see plots of land with short, spiky bushes sticking out of the dirt. They look like weeds, except for the few that have fluffy tufts of white cotton sticking to them. India produces around 24 million bales of cotton in a year and Vidarbha is the heart of this cotton country, right in the centre of the map. Although it isn't a food crop, cotton is one of the few plants that can survive the long, hot months here, so farmers in the area sell it to scrape a living while also growing small amounts of soybeans and *dal* between the fields for food.

Cotton-picking season is coming to an end. Some sun-wrinkled women are picking the last remaining balls, their saris hitched up around their knees in a sack to collect them.

We pass smallholding after smallholding, each burnt dry by the sun. There's a running herd of gazelle-like animals, which Jawandhia tells me are 'wild cows'. The only sign of modern civilisation is one little red tractor. 'There is almost no large-scale farming,' explains Jawandhia, sniffing into his sleeve. 'And that is why there is no machines until now. Even tractors were not here until six or seven years before. Now tractors is coming.'

People here are familiar with hard work, and disappointment. Only just over half of the farms are irrigated, meaning that thousands of families are at the mercy of the harsh weather. More than 15,000 villages in Vidarbha have water-shortage problems and the last soybean crop failed almost entirely because of a severe drought, forcing people to rely on their cotton earnings alone. The situation for some farmers is so severe that they resort to moneylenders who charge extortionate rates of interest. And this cycle of bad droughts and bad debts has driven hundreds of thousands of farmers to suicide.

So I find it strange that the houses in Kuljeri are painted a cheerful sunbleached turquoise. Surely all this misery should have somehow cast a shadow over their homes as well as over their lives? My car bumps through the tiny village before I run out of road altogether and grind to a halt. There's a barefoot, half-naked toddler wobbling through the dirt with a melting ice lolly in his hand.

Jawandhia and I walk through the village for a while, before we're invited inside the house of Ujawala Prabhakpetkar, a 38-

year-old mother who is also a small-scale farmer. There are goats chewing garbage outside her front door and, in the small open courtyard inside her home, among the wandering chickens, is a water well that has been drained empty.

'I don't like farming, but there's nothing else for me,' she says in Marathi, the local language. She has large, bright eyes and is wearing a saffron-coloured sari with a dip-dyed purple border. There are a few white plastic sacks of *dal* on the stone floor and a row of framed pictures of Hindu gods and goddesses above the entrance. We sit down over two small cups of sickly sweet tea in a narrow room between her kitchen and the courtyard.

About eight years ago her husband died, Ujawala tells me. She could have left the village, but decided to stay on the land with their two children. She tells me about her son and her daughter, neither of whom want to become farmers. And then there are her in-laws, she says, who are an extra burden that she wishes she didn't have. But at least they offer her some company as she grows older. After a while, her mood darkens and she tells me what caused her husband's death all those years ago. Her eyelids fall.

He was growing cotton on their small plot of land, she says, and borrowed cash from a moneylender so that he could grow four extra acres of soybeans, which were fetching a high price in the market at the time. Tragically, though, the crop was destroyed by a season of hailstones followed by unusually heavy rains. The family's debts piled up to around $4,000. Knowing that he couldn't pay back the money, Ujawala's husband became depressed and eventually took his own life. A silence falls over the three of us. She goes into the kitchen, drawing the folds of her sari over her face.

Jawandhia turns red with anger. 'See, this is what the tragedy is. If you go to cities like Delhi and Mumbai also, you see that in India we want cheap food, but we forget that the food producer is also living in this economy. That is why there is a gap between the rural and the urban people,' he says. Small-scale farming has always been a precarious way of life. Successive droughts back in 1966 even caused a famine in the Indian state of Bihar, in which thousands of lives were saved only because of emergency overseas food aid. But things are becoming tough again now as India develops.

'When I started in agriculture in 1970,' Jawandhia continues, 'the salary of a teacher was just 150 rupees per month. Today the salary of a teacher is no less than 15,000 rupees a month. A worker who was working on my field in 1970, I was giving just 100 or 150 rupees, which is same as the teacher's salary. Now the worker who is working on my field I cannot pay him more than 3,000 rupees a month. There is no more wages for the farm labourer.' With the wages of educated urban workers rising so much faster than those of farm labourers, rural life is becoming more desperate.

When a regional farmers' rights group started keeping track of the suicides they had been told about, they found that 241 farmers had committed suicide in this district – a region of 20 million people – in the first four months of 2010 alone. The year before, 916 people had taken their own lives. The same story has been repeated in other parts of the country, dating back to the early nineties. Between 1997 and the end of 2009, according to the National Crime Records Bureau, there were more than 199,000 suicides in rural areas.

In 1966, when famine loomed, Indian farmers were saved by science. Spurred by huge improvements in agricultural science all over the world, the government heavily promoted the use of pesticides, fertilisers, irrigation systems and high-yielding seeds.

These enormous changes worked so well that they became known as the Green Revolution. Food production swelled almost immediately. By 1979 India was producing such a high volume of grains that it was exporting the leftovers for the first time.

But by the turn of the millennium, another problem faced Indian farmers. The phenomenal gains brought about by the Green Revolution were tailing off. Factories couldn't produce chemicals any more powerful to squeeze any more out of the land, the old seeds had become as efficient as they were going to get, and pests were getting stronger by the day. So in 2002, farmers turned to science again. That was the year that the Indian government approved its first genetically modified seed for cultivation. Villages like Kuljeri finally got their magic beans.

Well, not beans exactly. India's first transgenic crop was a strain of cotton that was resistant to the American bollworm, a type of moth larvae that was until then voraciously munching its way through the country's cotton belt. Called Bt cotton, this variety was the product of years of expensive research by the US-based seed company Monsanto. They made it using a section of DNA from a bacterium called *bacillus thuringiensis*, which is toxic to many insects including the American bollworm. They took the exact gene responsible for this pest-resistant trait and popped it into the DNA of a normal cotton plant so that it would excrete the same toxins as the bacterium. Research like this is a hit-and-miss affair: new crop varieties often fail. But field trials showed that Bt cotton really worked. In fact, years before it arrived in India farmers had already successfully grown it in the US, China, Australia, Mexico, Argentina and South Africa. And it promised to be the first of a revolutionary new class of seeds in India, which would yield bumper crops every year.

Before Bt cotton came along, Indian farmers had no choice but to rely on the age-old system of breeding similar plants with each other to create stronger varieties. One type of rice would be crossed with another, hardier type of rice, for example.

Sometimes this process might be intensified and sped up by agricultural scientists in a laboratory, but it was still limited by the sorts of plants that would breed normally in nature – rice with rice and wheat with wheat, for example. Then genetic modification came along and suddenly it became possible to create plants with traits that nobody had seen before.

The new cotton was a hit. At first anyway. In 2006 a survey of Bt cotton farms in Maharashtra found that three quarters were bollworm-free, compared to the one third of those who weren't using the new seeds. Other pests, including caterpillars, were also down, which meant that farmers ended up using fewer pesticides. According to the Cotton Association of India, yields were up by more than 50 per cent by 2009. Even though it's difficult to say exactly how much of this was a direct result of farmers planting Bt cotton rather than other factors such as better weather, extra irrigation or pesticides, studies suggest that it certainly helped. Today, at least four fifths of all the cotton grown in India is the transgenic kind.

At the time, farmers dreamed of more unimaginably powerful plants, offering high yields, fewer pests and a miraculous resistance to droughts and floods, all thanks to the power of biotechnology.

Unfortunately, none of their dreams came true. Instead of a second Green Revolution, a battle broke out between the farmers and Monsanto.

It was all about the cost. Almost from the outset, Monsanto was criticised for the high price of its seeds. Newspaper reports claimed that its salespeople were also making exaggerated promises about impossibly fat harvests. Desperate farmers whose crops were being decimated by bollworm were led to believe that they had no other choice but to try Bt cotton. Although it worked for many, not everyone enjoyed great harvests every year. The seeds were not a panacea; they were simply a way to resist pests. Especially in dry regions like Vidarbha, bad weather and unhealthy soil sometimes conspired to give farmers less cotton than they had hoped for. Meanwhile, the price of the GM seeds stayed high.

Farmers across the country were furious. Small companies began illegally trading black-market Bt cotton seeds to meet the demand of farmers who couldn't afford to buy new ones from Monsanto every year. And anti-globalisation campaigners accused the American company of being profit-hungry and ruthless. In a few states, including Maharashtra, where I'm travelling now, the anger reached fever pitch, pushing local governments to force Monsanto to cap the price of its seeds to half their original cost.

But then the dispute hit a new low. Monsanto began to be blamed for causing even higher debts among cotton growers and driving more farmers to commit suicide than before. GM crops, activists claimed, were killing people.

In truth, there was no link between Bt cotton and the suicide epidemic among Indian farmers – or at least that was the conclusion of a comprehensive report by the International Food Policy Research Institute in 2008. Even so, the damage had already been done. The anger against Monsanto left a black mark against the technology. Since then people's fear of apparently greedy multinationals has become tangled up with a general nervousness about radical scientific research, especially if that research is foreign.

These days, people in Vidarbha don't just blame Monsanto for rising debt and farmers' suicides; they blame science itself. 'Science has increased the exploitation of the people. Technology is always used to exploit more and more,' says Vijay Jawandhia, the farmers' rights activist, adding that he's suspicious that Bt cotton might also be poisonous.

This has gone far beyond the safety concerns that surrounded the Flavr Savr tomato back in the US in the nineties. Memories of India's colonial past, when Western rulers exploited the poor, have also gone into the mix. The Minister for Farmers' Welfare and Agriculture Development in the state of Madhya Pradesh, which borders Maharasthra, denounced GM crops as a 'westernisation of agriculture science'. Some activists now advocate a return to traditional, old-fashioned farming without chemicals, fertilisers or modified seeds.

And this means that any prospect for a second Green Revolution is disappearing. Around a year ago, the Indian government announced its intention to introduce another GM crop to Indian fields, and debate is still raging about whether it should be allowed. The new crop is called Bt brinjal, a genetically modified version of the balloon-like purple vegetable known in the UK as the aubergine. Developed by Monsanto in partnership with an Indian seed company, it has the same pest-fighting properties as Monsanto's Bt cotton, but if it is given clearance for cultivation, it will be the first edible GM crop in India. Predictably perhaps, the response hasn't been positive.

Brinjal isn't a staple food in India but it is a popular one, used to make the famous spicy paste *baingan bharta*. For anti-GM activists, the idea of cooking a beloved national recipe with a transgenic vegetable created by an American company is tantamount to treason. Others complain that the new vegetable may not be safe to grow or eat either. In fact resistance to Monsanto's Bt brinjal has become so vociferous that what started out as a small battle between a multinational corporation and a few farmers has escalated into a full-scale war between scientists, politicians and campaigners.

Just recently, they went head to head at consultation meetings around the country. In the state of Orissa, where the government minister in charge of making the final decision about Bt brinjal turned up for a public meeting, hundreds of women organised a mock funeral procession and then burned an effigy of a Bt brinjal plant. In Kerala, 100,000 people reportedly fasted for a whole day as a mark of protest against Bt brinjal. Elsewhere, patriotic chefs prepared huge vats of *baingan bharta* in tribute to the brinjal.

Anti-GM and anti-Western activists have turned the whole issue into a matter of Indian sovereignty. They claim that the nation's farms are being overrun by expensive foreign technologies and that dangerous mutant seeds are displacing traditional ones.

The loudest cry of all comes from Vandana Shiva, a 57-year-old rural rights activist based in New Delhi and one of the world's most famous anti-globalisation protestors. In a masterful campaign over the past few months, she has passionately argued that irresponsible scientists are pushing an unwanted, potentially deadly technology onto unwitting and poor rural Indians. Shiva has been spearheading a crusade against the transgenic aubergine, harnessing the power of her supporters across India to lobby the government. Just recently she launched a silent march through New Delhi in the north, before travelling to Chennai in the south, where protestors held a cycle rally through the city.

Just recently, swayed by this apparently overwhelming tide of opposition, seven Indian states – between them growing 70 per cent of all of the country's natural brinjals – have announced a blanket ban on cultivation of Monsanto's Bt brinjal. In the end, as reported in the newspapers, the Indian government has had no choice but to put a temporary moratorium on the offending vegetable.

'The technology is stupid,' says Vandana Shiva. She has an unusual accent. A product of her travels around the world, it includes Indian and American but also sounds at times oddly Caribbean.

I'm at her farmers' rights NGO in New Delhi and the

atmosphere is frenetic. There are pictures of fat purple brinjals pinned all over the place, people are leaning over desks giving instructions and phones are ringing. This operation, belying its location in a converted house on a quiet residential street, is every bit as slick as those of giant global NGOs such as Greenpeace and Friends of the Earth.

Shiva is wearing a simple green sari accessorised with a thin gold chain and pearl earrings, with her grey hair scraped back into a bun. Her career as an international activist, she says, started after she earned her PhD in physics in Canada, where she wrote her dissertation on quantum theory. Then in the 1980s, she 'chucked in everything,' started an organic farm and became a global campaigner for ecological living. Her life transformed into that of an international jetsetter, with regular appearances on television networks around the world and cameos in leftwing documentaries alongside proponents of vegetarianism, organic farming and slow food.

Shiva is so powerful that, one year, a thousand tonnes of a corn and soya blend from two Christian charities, reportedly destined for Indian school lunches, was held up at a port in Kolkata while she lobbied the government not to accept it. There was a good chance, she claimed, that the American shipment contained GM food (more than half of all soybeans and corn grown in the US are GM varieties).

To this day whenever Shiva travels to the US, she tells me, she takes care to avoid eating anything that might have passed through a laboratory. It's not easy, since GM food is common in the US and generally isn't labelled, as it is in Europe.

'How do you do it?' I ask her.

'Well thank goodness I'm well known enough that people make an effort to get me good organic food from a local farmer.'

'What did you eat last time you were in the States?'

She laughs. Some young activists took her out one night 'dumpster diving' for dinner, she says, salvaging unwanted vegetables out of the trash bins behind restaurants and supermarkets. 'They are very clear, it's dumped on the day. They usually put out these tons and tons of salads or tons and tons of carrots, five days before the best before . . . it's totally edible and good.'

There are two other people in the office with us. One appears to be her secretary or assistant. He's a tall, thin, moustachioed man sitting at a desk at the other end of the room, watching me out of the corner of his eye. And the other is a skinny American or Canadian, no older than in his twenties. He looks as if he may be an intern or gap year student (Shiva has a dedicated following among young leftwingers in the West). The walls around us are lined with metal shelves containing huge stacks of pamphlets. *No GM Crops and Food! A Handbook for Activists*, I read on one cover. Another is titled *Genetic Modification and Frankenstein Foods*. Next to them is a row of books authored by Shiva, including *Stolen Harvest: The Hijacking of the Global Food Supply* and *Ecofeminism*.

One of the problems with transgenic crops, Shiva says, is that she believes they're unsafe because the genes inside strands of DNA are more complicated than researchers realise. Splicing a gene from one organism into another might not just pass on the trait you want; it could also lead to dangerous side effects that only emerge later on. They could include allergic reactions, for example, or even dangerous toxicity. There's also the risk that a GM plant could accidentally breed with a different one in the wild, spawning a new, mutant variety.

These are legitimate concerns of which, in fact, scientists have always been aware. On their advice, governments around the world have already taken steps to mitigate the chances of creating mutant crops and unforeseen allergies by heavily regulating their development. Everything is carefully measured, including how they are grown and the amount of time they need to be tested, and they go through trial after trial to make sure they don't cause side effects. Of the hundreds of transgenic crops that have been invented, only a handful of them have cleared these legislative hurdles. In India, this entire process is judged by a 31-member panel that includes nineteen expert scientists, known as the Genetic Engineering Approval Committee. Testing GM crops in rigorous field trials is such a hugely expensive enterprise and takes so many years, with no guarantee of a positive outcome, that the cost of meeting these tough safety standards is partly why Monsanto charges a premium for its seeds.

'Aren't you reassured by all this regulation?' I ask Shiva.

'No!' she says, clutching a mug of tea and staring intently into my eyes. 'We're killing our farmers. Is the farmer earning enough? No, the farmer is spending more than they're earning! You cannot be financially sustainable with that. Are we using water in a sustainable way? No we're not, we're using ten times more water to produce the same amount of food! Are we sustaining the climate? We're not! And are we sustaining biodiversity? We're not!'

'Do you think that, given the chance, farmers would ditch their Bt cotton seeds and go back to ordinary cotton?'

'Yes, absolutely, I'm sure,' she says. 'The only reason they can't, she argues, is because years of cultivating transgenic cotton has meant that the old seeds have been irretrievably lost.

She's towering and loud. I'm reluctant to argue with her.

The solution to the problems faced by Indian farmers, Shiva continues, is to go entirely organic and grow all crops without artificial pesticides, fertilisers or the help of laboratory scientists. 'My argument is that India is too heavily populated and with too little resources, with 70 per cent of our people dependent on agriculture, that we cannot play games with our ecosystems. Right? And I repeat again and again and again, it is crude tech. It is stupid tech,' she says, finally dropping her voice to a near whisper. 'What drives me is that lies are told. I hold science so dear to my heart, I can't tolerate lies.'

Shiva's radical beliefs go far beyond opposition to just transgenic vegetables; she's against any kind of scientific interference in farming. The Green Revolution – the first round of technical advances that helped India beat famine back in the 1970s – was a global conspiracy, she says, to make farmers believe that they needed the help of new technology to improve their yields. 'It was totally unnecessary, totally unnecessary,' she says. 'Because we've done the calculations. And it's the additional land and the additional water made available to those two crops – rice and wheat – that accounts for the increase in rice and wheat production. It's been falsely allocated to miracle seeds and miracle chemicals. You give that kind of land and water for indigenous seeds, with organic farming, you'll have the same amount of increase.'

It's difficult to prove whether or not this is true. The general consensus among politicians, economists and scientists is that the Green Revolution was not only necessary, but that it was a success. The American agricultural scientist the late Norman Borlaug, who introduced high-yielding seed varieties to India, Pakistan and Mexico, was even awarded the Nobel Peace Prize in 1970 for his efforts.

Shiva knows that her views put her in a minority among fellow scientists, but she considers herself a maverick hero, challenging the 'stupid' system.

By her own admission, her role model is Mahatma Gandhi, the tireless freedom fighter who was also, sixty years earlier, a champion of India's self-sufficient, small-scale farmers. One of Gandhi's beliefs, underpinning his peaceful mass movement, was that families should be encouraged to grow and eat their own food and thereby undermine the power that British colonisers had over them. It wasn't only a rejection of foreign rule over India, it was also a rejection of Western modernity and ideas. Gandhi's legacy is so profound that, to this day, local governments in Indian states have legislation that caps the amount of land a person can own, to make sure that citizens don't lose their farms to giant corporations or rich landowners. To keep the ceiling in place, Indian leaders have redistributed land to millions of poor families. This is the reason why most Indian farmers own only a few acres, forcing many to live almost subsistence lifestyles. In Maharashtra, for example, Vijay Jawandhia is considered to be a large-scale farmer because he has fifty acres, which is just under the limit legally allocated to him. This is tiny compared to most farms in the West.

Just as Indian politicians still wear clothes made of homespun cotton to demonstrate their simple, Gandhian values, Shiva has adopted a similar image. Her unform is a plain sari and a simple *bindi*. Her writings celebrate India's traditional villages and call for citizens to preserve the nation's natural ecology and indigenous resources. It's all about a return to traditional values and a traditional way of life. And it puts modern, Western science and American companies like Monsanto up against poor and vulnerable Indian farmers.

'Gandhi has been a tremendous guide in finding a way to walk when there seems to be no opening,' she says.

'In what way?'

'Well, there is a lot of violence in farming on the industrial model.'

'Violence?' I have no idea what she means.

'Fertilisers come from explosives factories . . .' she begins to explain.

By this I'm assuming she's referring to the fact that ammonium nitrate is a fertiliser and also an ingredient in TNT.

'. . . Herbicides were also war chemicals,' she continues. 'So all of this violence, which was designed to kill in the war, when it's employed in agriculture it creates a similar warlike situation against the species, and finally also against the farmers themselves.'

Shiva seems to be drawing parallels between Gandhi's peaceful struggle against colonisation and her own 'non-violent' attitude to farming. But if there are dangers in transgenic crops, Shiva's alternative isn't without its own pitfalls too. Her vision of a land of small, organic farms is also flawed, according to some scientists and modernisers. Having basic smallholdings means that most Indian farmers can't afford machinery like tractors, artificial irrigators or climate-controlled food stores. Large-scale, industrial farming of the type found across Europe and the US is close to impossible.

Nonetheless, her emotive rhetoric is working. Because of Gandhi's legacy and powerful activists like Shiva, the idea of small-scale land ownership and traditional ways of cultivating this land has become integral to the country's vision of itself. For people like Shiva, India is not a nation of geeks; it will always be a nation of farmers.

This exploits a natural division between India's traditionalists and the country's modernists. It also cuts out any prospect for using new science, like that practised by the National Botanical Research Institute, on India's small farms.

The strange thing, though, is that Shiva is proud of her physics degree. In fact, it was one of the first things she mentioned when I arrived. And this makes me wonder whether her scientific training has ever affected her convictions that organic, small-scale farming is right for India, or whether, contrary to what I first assumed, studying physics has actually reinforced her beliefs.

'It's not quantum theory, but I think a lot of my quantum theory training helped me in the systems analysis that I do,' she says.

'In what way?'

'Well, quantum theory teaches you that things are connected. And that's why, for me, genetic engineering is a double tragedy, because it's using a paradigm of a mechanistic, reductionist science, which we threw out of physics a century ago, 1905 onwards. In biology these things are so obvious, that this plant wouldn't survive without this soil, you know separating soil from plant and plant from pest, that everything is interrelated.'

Again, I don't understand Shiva's argument. I think she's somehow applied physics, particularly the idea of quantum entanglement (which says that each particle in the universe is connected to every other, however distant it is, on a fundamental level) to her own philosophy about how plants and pests should not be separated because they are part of an interdependent ecology. She's mixed two unrelated branches of science. At least, that's what I think she's done. I'm confused. There's something mystical about her beliefs, reminding me of Hindu ideas about cosmic consciousness and Buddhist philosophies about the interconnectedness of the universe.

'Genetic engineering is based on the idea of immutable atoms, but genes are not immutable atoms,' Shiva continues, again mixing physics with biology. 'In the final analysis there are no immutable properties. There are interconnections and possibilities and potentials.'

'Do you think genetic scientists have anything useful to offer farming at all?' I ask, in an attempt to drag the conversation back to useful facts.

'In terms of science, they don't do science, they function as technicians,' she says, dismissively. 'You know, a technician using an existing tool. A scientist asks questions and answers them and tells you something new about the world. They can't tell you how a plant functions, they can't tell you how pests really get controlled.'

'So technology has no place in farming?'

'If I want to put this painting up and what I want to do is put a nail into this wall, I bring a hammer,' she says. 'That is the appropriate technology. But you say, no, I have a more sophisticated instrument. I have an earth-moving machine next door, and you try and bring it, and say, I'll fit it with that. Are you being stupid or smart? Tell me. Go on, tell me.'

She stares at me, waiting for an answer. The man sitting at the desk at the other end of the room is watching me too. 'Stupid?' I reply, finally.

Shiva flashes me a smile. 'The size of the machine, the sophistication of the machine, the high-tech nature of the machine doesn't make the application high-tech. And that's the point I want to make. The ability to shoot genes across species barriers is like the earth mover, but it's just the wrong way to do breeding,' she says.

Many scientists disagree.

Among them are the molecular biologists working in the National Botanical Research Institute, where research into the long-life banana is happening. The institute is in the historic city of Lucknow, in the state of Uttar Pradesh. And it's one of the most productive labs in the country. In the last year, they published 172 scientific papers in national and international journals and were granted twenty-one patents. Home to some of the nation's most respected agricultural scientists and geneticists, Lucknow has developed a reputation as an emerging biotechnology hub.

Outside the laboratories, though, parts of the city are a mess. Like India's capital, Lucknow's pink and yellow palaces date back hundreds of years to when rich and powerful Muslim nobles governed the area. But now parts of the city feel almost Dickensian. The wider state of Uttar Pradesh is among India's poorest, yet the newspapers recently photographed the chief minister having a giant garland of high-value banknotes draped around her neck by her political followers. I find myself stuck in traffic for an hour when she returns from a trip and the streets are being cleared for her arrival. I stare out of the car window, my forehead pressed against the glass. The dirty streets are lined with lopsided redbrick shacks. There are begging mothers, babes in arms, and row upon row of wobbly cycle-rickshaws with torn plastic seats and rusty spokes.

Amidst this chaos, the gardens and laboratories of the National Botanical Research Institute are an oasis. For a change, the outside world feels nuttier than the professors. Hidden behind iron gates, carefully tended beds of plants and flowers fill a giant circle, bordered by curving paths and meticulous lawns. On a nature trail through the heart of the institute, I pass tall white and pink gladiolus, bougainvillea, chrysanthemums and small orange marigolds. And at the end of this path, I arrive at the Plant Transgenic House, a giant greenhouse painted sky blue with double glass doors. Behind the doors are long rows of tomato plants, cotton and chickpeas. All of them are genetically modified. Every room is a trove of transgenic wonders.

Just recently, scientists here at the National Botanical Research Institute completed work on their own domestic version of Bt cotton, which an Indian seed company has been selling as a rival to Monsanto's seeds for about a year. The work was done just above the long-life banana laboratory, where the same group of researchers is now working on other crops to make them similarly resistant to pests and viruses. They've already invented a Bt chickpea, a tomato plant that's less likely to catch the genetic plant disease geminivirus, and a cotton plant that won't be eaten by the cotton leafworm, says Dr Pradhyomna Kumar Singh, who works in the team. He's a stocky guy with a lisp, wearing a stripy orange shirt.

'Monsanto, they have huge infrastructure, hell lotta money. They spend money like anything and they can afford all kinds of experiments,' Singh mutters (a little bitterly) as we walk into his office on the next floor. But even though Monsanto got there first, his team is achieving remarkable results. Their seeds are cheaper than Monsanto's seeds. And already about a tenth of the Bt cotton seeds planted on Indian farms are homegrown varieties, including the one developed here.

'Does Bt cotton work?' I ask him, with Vandana Shiva's comments in my mind.

'Yes. It is a very, very powerful tool. The kind of productivity, the kind of crop protection we achieved with Bt cotton, it was not possible *ever* before.'

'What about Bt brinjal? Is it safe to eat?'

'I think Bt brinjal is safe, as of now. They should be released, I believe,' says Singh. He believes that the negative public reaction to GM crops in India has largely been because foreign corporations are selling their expensive seeds to the poor. Had Bt brinjal been developed by an Indian company or by Indian government scientists like him instead of an American multinational, there would have been far less protest, he says.

Perhaps because of the concern that American companies like Monsanto have too much control over the GM seed market, the government has pumped extra money into developing domestic transgenic crops. Politicians hope that these varieties will be more popular with the public than Monsanto's Bt cotton and Bt brinjal. And if they are successful, they will also instantly overcome the main problem of Monsanto's Bt cotton – they will be cheaper because they're made in India. The National Botanical Research Institute already falls under the banner of the government-funded Council for Scientific and Industrial Research, but it has support from the highest levels, including the prime minister, Manmohan Singh, who has been a vocal supporter of GM technology.

Encouraged by the government's backing for transgenic crops, downstairs, Pravendra Nath is extending his research on the long-life banana to other fruits, vegetables and flowers. 'Fruit we are working on now is the banana and mango, vegetable at the moment we're using tomato as a model system, and in flowers we got rose, gladiolus,' he explains. There's a blow heater in one corner to ward off the last chilly winds of the winter and keep the small shoots warm. In the middle are three long laboratory benches, stuffed with paper files and glass flasks of all sizes. Half a dozen young scientists in white coats are taking measurements and poring over petri dishes.

I ask him how his work on the long-life banana is going. For the last seven years, he says, the nineteen researchers in his team have been painstakingly looking through old scientific papers to see which genes inside the banana are responsible for different traits, comparing them to other fruits, and then going through each one. They have been testing thousands of genes in arrays, activating and deactivating each one in turn. To some extent, this is all a matter of trial and error. If you are lucky, he says, it can take only two or three years, but more often it takes longer. It takes patience to be a geneticist.

Nath shows me the latest results. He calls up a set of pictures on his computer screen from the most recent tests carried out on their new banana. The document is titled 'Delayed ripening in transgenic bananas'. There are photographs of two fresh bunches of bananas on the screen. He clicks. The next picture was taken five days later and shows how one of these bunches has turned black, while the other is still bright yellow. He clicks on the laptop's keypad again. After ten days, the first bunch has gone uniformly black and looks inedible, while the second – the genetically modified bunch – has developed only a couple of black spots and is still good to eat.

'How much longer does your banana last?' I ask.

'At least three to four days extension is there,' he says, smiling. It is something only one other research team in the world has achieved, as far as he knows. A group at Cornell University in the US recently got similar results with their own banana, but as yet neither of them has published their results.

Nath's team has replicated their banana research on tomatoes.

He shows me another set of pictures, these ones entitled 'Increase in on-vine and post-harvest life of transgenic tomatoes.' The four tomatoes at the top of the photograph began to blister after nineteen days, while the tomatoes underneath took thirty-one days to do the same. There is at least a week's difference between when the ordinary tomato needed to be harvested and when the transgenic tomato needed to be harvested. 'What we found was that, by using this technology, the growth of the tomato, the shape, size, taste and other parameters are the same. They mature and ripen almost at the same time, but our ones stay for a longer time on the vine after they have ripened,' he explains. 'The normal tomato it will just drop after three or four days, but this tomato, our experimental tomato, this will not drop even after ten days. So the advantage for the farmer is you can harvest your crop in small portions. We're working with this objective that it should revolutionise and we should be able to save a lot of fruit and a lot of money.' Nath expects that it will take another three years before these fruits make it into Indian kitchens, because of the stringent testing and regulatory hurdles.

So before I leave his lab, I ask him whether transgenic crops like his long-life tomatoes and bananas can really solve the problem of crippling poverty among Indian farmers. 'I wouldn't say GM crops are the only way,' he admits. 'But at the moment, when you're racing against the time, then I think GM crops are the solution. I foresee that GM food *will* come out in India. The scientists in the country, they're pretty sure about their achievements and the good or bad points of their research. And I think they have been able to convince the government that GM food is not as bad as it has been publicised.'

Dr Ajay Parida, the 46-year-old executive director of the M S Swaminathan Research Foundation – another Indian institute that does research into transgenic crops – agrees that GM crops are vital for Indian farms. He's a large man with a bushy grey moustache, and he's had a front seat in the entire debate: his boss, Professor M S Swaminathan, was the scientist who helped design India's Green Revolution forty years ago.

Today the foundation works closely with farmers across India and has a big influence on government policy. It has a small training centre, fitted with a computer, near the village of Kuljeri in Vidarbha, from where it instructs locals on new farming techniques. The style of the foundation's headquarters sits somewhere between scientific modernity and down-to-earth Indian style. Small laboratories circle a traditional courtyard containing tiny bunches of labelled grasses, trees and flowers, like a kitchen garden. There's a crop of aloe and another of yellow bamboo.

Parida is a no-nonsense man. The plight of Indian farmers is an emotional issue for activists and politicians, but he, by contrast, treats it as a scientific problem that can be fixed. Although they may seem to be anti-science, he says, most farmers tend to have a problem with Monsanto and its practices rather than with the technology underpinning transgenic crops. 'If a public-sector research institute had developed this particular technology, it would have been entirely different,' he says. 'GM is very essential. So for example, you have questions of stress. In India, 62 per cent of the cropping area is rain-fed, so what you require then is crop species that are responsive to these kinds of needs. What you require in the long run is drought-tolerant species, while some places require water tolerance because of floods and things, and most of all, heat resistance, because the temperature is also rising because of climate change. Where will all these things come from?'

Rice, for example, is one of the country's staple foods and the demand for it is outstripping the space available to grow the crop. So Parida and his colleagues are developing a strain of rice that's able to grow in the salty soils found along India's coastal plains. They're doing it by splicing genes from plants and trees that grow in the mangrove forests in the seawater areas where south India meets the Indian Ocean, and inserting them into rice DNA.

'What we did first,' he explains, 'was we took the basmati rice, because that is good system for the research purpose and put our transgenic material into basmati. And we have done the same with a *ponni* variety [eaten in the southern state of Tamil Nadu, where people prefer to eat parboiled rather than boiled rice] and several local varieties.'

They've started some limited field trials, Parida explains, and it will take another two or three years after that before the salt-tolerant rice will go through the same process as Bt brinjal – a long government consultation to decide whether the public are happy for it to be grown in India. It's about the same amount of time it will take for Pravendra Nath's long-life bananas and tomatoes and Pradhyomna Kumar Singh's virus-proof chickpeas to reach a similar stage. In fact, Parida tells me, soon research into transgenic crops will have matured to the point where scientists like him will be able to transform Indian farming. Research and development in Indian laboratories into about twenty-five promising GM varieties will be concluded in the next few years. When that happens, there will be enough transgenic vegetables and fruits to fill a kitchen. They include types of rice, okra, cabbage, potato, corn, cauliflower and sugar cane. If the government approves them all, these crops could together help to bring about a second Green Revolution.

But the Bt brinjal controversy makes me wonder if this is actually possible. Anti-GM activists, farmers and some Indian politicians have become so ideologically opposed to the underlying technology that it may be too late for scientists to convince them otherwise. So I ask Parida what would happen if Indian farmers were to ultimately reject the work of scientists like him and instead go the other way, by returning to traditional or organic agriculture, as Vandana Shiva advocates.

He looks up from the notes on his desk and pauses. They won't produce enough to feed the population, he says.

Before I leave, Parida offers to show me the gardens around his laboratories. We pass a small tree with golden, almond-shaped leaves. In a lonely corner of another flowerbed is a weeping fig.

'The place where we are standing now, ten years back or fifteen years back, the surrounding area was completely paddy fields. It was all agricultural fields,' he says, staring nostalgically into the distance. 'If you look now, it's all the high-rising software companies.'

Past the fig is a row of office blocks stretching into the horizon. India seems to be changing. And the statistics back this up. By 2030 as many as half a billion people could be living in India's cities, compared to around 340 million now. More than two thirds of new jobs will come from these urban areas. And as the cities grow, traditional villages are already turning into anachronisms. 'Farmers are losing interest in doing farming as a profession, and young people are not coming up, taking farming as a profession,' says Parida. 'Even if you go down the road here and see the software companies, you will find all the security guards and all the maintenance staff in these high rise buildings are rural people from Bihar, Jharkhand and Orissa, who would have worked in farming.'

Mahatma Gandhi's vision of a nation of self-sustaining farming families is crumbling away underneath the heavy weight of economic and technological change. In some ways, I can't help feeling that it's inevitable. There isn't an industrialised country on earth where the majority of the population still works on the soil. If India dreams of joining these ranks and becoming a superpower, then village life will inevitably give way to sprawling cities, giant factories and large-scale industrial agriculture. One day the bullock carts will probably disappear and tractors will replace them.

Whatever the outcome of the war over GM foods, the heart of the problem is that farmers are struggling to adapt to the new India. They need higher yields and better crops to feed the growing masses and pay for their changing lifestyles, but they're struggling to find a way forward in a society that treats them as though they still living in the early 1900s. What's confusing me now is that, when the alternative to their low-yield, traditional farming is migration or suicide, why do Indian farmers maintain such an ideological opposition to transgenic crops, as Vandana Shiva claims they do? Why do farmers who have been so desperate to grow Bt cotton in regions like Vidarbha have a problem with other GM crops?

'Is it that farmers distrust science?' I ask Parida.

No, he insists. Farmers can see how useful science can be. 'If farmers do not see a benefit of this kind of technology, they do not go for the technology. I mean almost all of the cotton grown in this country is Bt cotton. All of these people growing Bt cotton are small-scale farmers, and they're doing it despite the high cost of the seed,' he says.

This makes me question whether the opposition to transgenic crops is as high as anti-GM activists like Vandana Shiva believe.

Back in Ujawala Prabhakpetkar's small turquoise home, some neighbours join us. 'How have the harvests been?' I ask everyone.

'The field is not good here. We will go to the city to work if we can,' complains Vinod Thackeray, a small-scale farmer growing

Bt cotton on fifteen acres of land, who lives opposite her.

Ujawala agrees. Her 17-year-old daughter has already left for the nearby city of Nagpur to study nursing, and her 13-year-old son dreams of becoming an engineer.

'Won't you miss them?' I ask her.

'It's too late for me, but I want my children to get an education and leave,' she replies simply.

The lightning pace of migration from the country to the cities means that this may be one of the last few generations to choose to work on the land. Makeshift roadside colleges in this area offer cheap IT courses for people who want to leave the villages and work in call centres. In the car on the way here, I gave a lift to a teenager who told me she had an exam to sit in the nearby town of Waifad.

Looking out over the spiky cotton fields outside Ujawala's home, the sun burning the last white tufts of the season, I recall what Vandana Shiva had told me when I visited her anti-GM campaign headquarters. She had said that, if they were given the chance, farmers would ditch Bt cotton and return to growing ordinary cotton, like they used to. I want to know if this is true. 'Would you ever go back to growing cheaper, ordinary cotton instead of the GM kind?' I ask Thackeray.

'Of course not!' he says. If he did, pests would eat up his crop within a season.

Perhaps then, Shiva was wrong – or at least as far as this farmer is concerned.

Then Vijay Jawandhia, the vocal farmers' rights activist who had been so vociferous in his attacks against science and technology, also reluctantly admits that he has started growing a domestic strain of Bt cotton, of the kind developed at the National Botanical Research Institute in Lucknow, on his fields.

'How is it doing?' I ask him.

He looks embarrassed, as though he's been accidentally caught out. 'It was not very good and not very bad also,' he says. 'Because you know, this year we're having a very severe drought condition in this area. The soybean crop is totally failed and cotton crop is also successful on only twenty per cent of lands. I think that variety is not that bad.'

'Why did you decide to grow it at all?'

'Because . . . see . . . I am a common man. I don't know anything,' he says sheepishly.

'I know that's not true! You told me you had a degree in chemistry.'

He smiles. He's growing it because it's an Indian variety and so he trusts it, he admits.

Indeed, Jawandhia has been so impressed by Bt cotton, he tells me, that he hopes Indian scientists will develop other GM crops like this one, which take into account concerns about price and local taste. 'It is really a tragedy that government is not pushing their project of introducing such types of varieties more and more,' he says, to my surprise. 'For so big country with so many agricultural scientists, why there are not twenty varieties, or forty varieties?'

In this small farming village in Vidarbha at least, it feels as though the big battle over GM crops, between the corporations and the activists, and the scientists and traditionalists, is over without either side even realising it. Whatever Monsanto has done to promote its seeds, whatever Vandana Shiva has done to propagate her vision of a traditional, small-scale, organic nation, and whatever protests and debates may be happening in India's cities, in the end, farmers here want to put food on their tables. If transgenic crops improve their harvests and if they can afford them, then perhaps there's no argument powerful enough to stop them from planting the seeds.

CHAPTER 4

CHARIOTS OF THE GODS

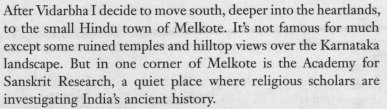

After Vidarbha I decide to move south, deeper into the heartlands, to the small Hindu town of Melkote. It's not famous for much except some ruined temples and hilltop views over the Karnataka landscape. But in one corner of Melkote is the Academy for Sanskrit Research, a quiet place where religious scholars are investigating India's ancient history.

This leg of my journey brings to mind that tired foreign-travel-guide cliché that India is this land of contrasts, of the modern and traditional, the spiritual and scientific. Maybe it's the only way to explain somewhere so vast and ancient that it naturally defies explanation, but there is some truth to the fact that, whatever change happens here, India never completely relinquishes what went before. Religious myths can live on in the popular culture as vividly as the stories on the front pages of the newspapers. Like the travel guides say, and as Vidarbha has already proved, the old sits alongside the new.

And so I'm hoping that by understanding its past, I might be able to get a better grip on India's geeky future.

In the early twentieth century, a mysterious holy man in south India wrote a scientific paper. By all accounts – although there aren't many of them – he was an unlikely scientist. He was born into poverty, had no schooling, spent most of his childhood begging and later survived smallpox. As he grew older, he adopted the same obscure life of asceticism as thousands of Hindu hermits across the country. On the surface, he probably looked as though he knew nothing about modern science. Yet in his paper, he suggested that the Vedas, which are Hinduism's oldest scriptures, comprising mantras dating back at least 3,000 years, contained the blueprints of a hitherto unknown technology used by early Indians and ancient gods. He had decoded these religious texts by channelling the minds of the deities, it seemed, and figured out how these machines worked. He called his manuscript, written in 6,000 lines of verse, the *Vaimanika Shastra*.

For a few decades, however, it was pretty much ignored. The holy man died and his work was forgotten. But then in the 1950s, a Hindu academic called G R Josyer stumbled upon it. At the end of the decade, he published a painstaking modern-language Hindi translation of the *Vaimanika Shastra*. Then in 1973, he translated it again, this time into English. The book circulated around the world, its title sending ripples of excitement wherever it was read. It was called *Science of Aeronautics . . . All About Machines*.

People already knew that Indian religious scriptures talked of ancient warriors who travelled in floating vehicles (known as the mythical 'chariots of the gods') but most people assumed that this was just fanciful storytelling or allegory. Josyer's book suggested there might be more to the stories than fiction. He declared that the *Vaimanika Shastra* was not just another scrap of philosophy of the kind routinely written by holy men, but that it contained descriptions of real aircraft that had existed thousands of years ago. He went to the trouble of giving technical particulars and included detailed blueprints by an engineering draughtsman from Bengaluru.

On the baby-pink cover of his book was a small pencil drawing of one of these planes. Part submarine, part mechanical fish, it was built in four tiers like a wedding cake, with three fins and a thin propeller at the front. Among the substances powering this unlikely contraption, the book said, was mercury, the silver-coloured liquid metal used in thermometers. Other ingredients included snake poison, rhinoceros bones and camel urine.

On other pages, underneath pictures of hovering babies carrying garlands of flowers, were cross-sections of other flying chariots. One was drawn in beautifully exact thin green lines, illustrating the position of air heaters, blowers, a driving shaft, a chimney and huge feathers stretching out along the back. It looked like a mechanical bird. In fact Josyer described it as a plane 'which can fly in the sky with speed equal to that of birds'. On another page was an aircraft shaped like a cone with the top sliced off, air pipes and electric magnets skewering its length. There were three small propellers at the top, two platforms for carrying passengers and one for the pilot.

And there was more to Josyer's book than just drawings. It also included lengthy explanations of the personal qualities needed by the pilots ('he must know the structure of the aeroplane, know the means of its take off and ascent to the sky, know how to drive it and how to halt it when necessary, how to manoeuvre it and make it perform spectacular feats in the sky without crashing'). There were recipes too, for what these pilots should eat ('56 [roots] should be purified, powdered, and duly cooked, and made into balls, and given out for use as food'). And there was a detailed description of how to generate electricity to power the dynamos that would drive the aircraft ('get a . . . flame-faced lion's skin, duly cleaned, add salt, and placing in the vessel containing spike-grass acid, boil for . . . 15 hours. Then wash it with cold water').

The foreword to Josyer's book described the holy man who originally wrote the *Vaimanika Shastra* as 'a walking lexicon gifted with occult perception. His sole aim was to transmit his knowledge to posterity. He lived a life of poverty, like Socrates, and sought no gains for himself . . . The 20th century may be said to be made historic by two achievements: the bringing of Moon-rock from outer space, and the publication of *Vaimanika Shastra* from the unknown past. [It] is a Cornucopia of precious formulas for the manufacture of Aeroplanes.'

Josyer's translation attracted letters from Sweden, Italy, Germany and the US. He even got an invitation to tea with the Maharaja of Mysore. Some people saw this esoteric old manuscript as evidence of advanced ancient Indian civilisations. Others claimed it as proof that aliens from outer space had visited India thousands of years ago (the *Vaimanika Shastra* appears on UFO websites even now).

In the late 1970s the local government in Karnataka decided to give G R Josyer a plot of land in the small town of Melkote on which to start the Academy of Sanskrit Research. The idea was that he and a team of scholars would investigate the science and technology of the Vedas even further.

There are some places in the world that feel as if nobody else has been there for decades. Even the people are frozen in time. The Academy of Sanskrit Research, on top of a steep hill in the town of Melkote, is one of those places. It's only a few hours' drive from Bengaluru, but the smooth, black Tarmac road that takes me there is so empty that farmers are using it to spread out their hay so it can dry in the sun. It crunches when I drive over it. Further along the road I see ten people squeezed into a rattling rickshaw, and then a herd of small black goats.

If you consider India's past, it's easy to see why this is still the most religious country on earth. It's the birthplace of four of the world's major faiths, Hinduism, Buddhism, Jainism and Sikhism. Even now, the percentage of atheists is in single digits. Here, the past lives on in religious tradition and superstition. It's normal for respectable politicians and billionaire business owners to consult swamis and wandering sages for good luck. Farmers, like the ones I met in Vidarbha, turn to astrologers to tell them their future, and so too do scientists in cities like Lucknow.

But how does this all square with India's geeky future? In other parts of the world, religion and tradition have come up against science and clashed badly (in Europe, for example, the number of regular churchgoers has steadily fallen over the years), yet in India, there's no sign of the same happening. Of course, India has a legacy in mathematics and astronomy that dates back thousands of years – the ancient mathematical text, the Bakhshali Manuscript, is just one example.

And a great deal of traditional knowledge, especially that related to health, has also survived the centuries and become absorbed into everyday Indian life. Parts of it are so scientifically sound that they have also entered modern science. Pharmacologists working for the world's biggest drug companies, for example, have developed useful medicines from old Indian herbal treatments. Neem tree extract, which Indians have used for at least two millennia in toothpaste and soap, is a proven insect repellent and fungicide. And turmeric, a yellow spice liberally used in Indian cooking, is being investigated in the US as a potential treatment for Alzheimer's disease and cancer.

But the problem lies at the far end of the spectrum, where science seems to be confused with legend and myth; and people continue to believe that manuscripts like the *Vaimanika Shastra* are literally true.

The academy, I hope, might have some answers. It sits at the far end of town, past low brick shacks, trees shaped like coat hangers, a line of ice-cream vendors and the beige stone ruins of an old Hindu temple. The whole place is bathed in sunlight and silent, bordered by red and pink hibiscus flowers. In one corner is a pair of peahens basking in a seven-feet-high metal birdcage. Opposite them is a blue solar panel that helps top up the Academy's electricity supply.

A man dressed in long robes and a beard asks me to take off my shoes before I enter the academy. I hop over the sunbaked stone floors as he guides me to the scientific research wing. Here, one of the oldest experts in the academy, Subbarao Narayana, is waiting to greet me.

The scholars in this academy, I've heard, believe that the *Vaimanika Shastra* is just the tip of the scientific iceberg. Hinduism's oldest scriptures: they say, contain the secrets of everything, like encyclopaedias of the universe. They point to the title of the Hindu scriptures: the meaning of the word 'veda' is 'knowledge'.

Narayana, a short man wearing a cream-coloured shirt and a white sarong, has worked at the Academy for twenty-five years, trying to dig this hidden information out of these old texts. He speaks such flawless Hindi that I struggle to understand him. I'm used to Hinglish, a mix of Hindi and English. 'I'm here to find out more about science in the religious manuscripts,' I say, in my imperfect Hinglish.

He attempts some English for my benefit. 'You are right to come. The information is there in the texts, going back 5,000 years. There is special knowledge in them,' he replies, pointing to a bunch of folded palm-leaf manuscripts on the shelf next to him. 'Most of them, they don't believe it,' he says, with a pitying smile. 'They don't understand that it's all science, everything around us – the lights, the computer, everything. But the ancient knowledge is inside us. These scientists need to study the Hindu

Vedas to get the understanding. One who wants to know the knowledge of the Vedas, he should go to a guru.'

'And how do the gurus unlock this hidden knowledge?' I ask.

'Gurus are the ones who have written about scientific knowledge, and they got their information by meditating and by intuition.'

The problem with decoding religious manuscripts, he tells me, is that their meaning is so carefully hidden behind metaphors, verse and myths that it's difficult to decipher. It's a task made even tougher by the fact that few people nowadays read the ancient language Sanskrit. Scientists rarely take an interest in the holy scriptures, he complains.

The *Vaimanika Shastra* is a rare exception. And so scholars here at the Academy of Sanskrit Research have spent decades building on Josyer's original work on flying machines. They now have proof, Narayana says, that the aircraft were built using a mix of mysterious metals and other ingredients that were invisible to radar, so their enemies couldn't find them. There is more, he adds, peering over a pile of papers on his desk. 'How to create the plane used by the saints 5,000 years ago, the fuel was a solar engine.'

'They used solar power?'

'Yes,' he says, digging through his books, trying to find the relevant parts of his work for me. He can't find them. 'If you want I can show you what the aircraft looked like?' he asks finally.

There is an entire room here dedicated to the *Vaimanika Shastra*, at the end of the corridor near his office. On the way we pass through the library, where a young woman in an orange sari is sitting crosslegged on the floor writing notes. On one shelf there is a copy of a book entitled *Vedic Mathematics for all Ages*. At the end is an old-fashioned mechanical printing press, which slowly churns out the leaflets and booklets written by the people who work here.

When we reach the room, Narayana opens the door with a

proud smile. On one side, opposite a couch and some crowded bookshelves, is a row of display boards with pictures of aircraft and quotes from the *Vaimanika Shastra* pinned to them.

'What do you think?' he asks, raising his eyebrows.

'Let me have a look.' There is one drawing of a solar-powered aircraft that looks like a blimp with wheels. Next to this is a row of toy aeroplanes, including a tiny black stealth aircraft and a yellow and red fighter jet. I scan further along. A colour painting on another display board shows a group of warriors in suits of armour standing in a field, pointing to a gigantic disc hovering in the distance. It looks like a flying saucer.

Everything gives off the unmistakeable whiff of crackpot.

'It can operate underwater, on land and in space,' Narayana reads for me from the wall.

'Space? Do you mean that people travelled into space? Outer space?' I ask him. I study Narayana's face, struggling to understand how a man of his age and education could believe that early Indians flew around in aeroplanes that looked like flying saucers.

'Yes, of course,' he replies. 'They travelled at . . .' He struggles to remember the right word. '*Supersonic* speeds. They even went to different planets.'

'Other planets?'

'They got there using their divine power, but we can't do that now. It's very difficult.'

'And what happened on these planets?' I ask.

'They met with aliens,' he says, looking me square in the face.

Religious faith can be a powerful thing.

Subbarao Narayana, though, isn't the only person who is convinced that Indians built flying machines 5,000 years ago and then rode them to meet aliens on other planets. Although it's hard to know exactly, the fact that the Academy of Sanskrit Research continues to attract governmental support suggests that thousands, maybe even hundreds of thousands, of Indians do. In my journey so far, I've even met professional scientists who believe that universal knowledge is contained in the Vedas. And this has left me with a new question. Is it just religious faith, or is there something deeper in Indian society, that makes people believe that pseudoscientific manuscripts like the *Vaimanika Shastra* are literally true?

The best way to find out is to look further into the country's past. So now I'm headed for Mysore, a city not far from Melkote, which is home to one of India's oldest libraries. The famous Oriental Research Institute is a giant repository for ancient texts, some of which date as far back as the fifteenth century. If there are any more clues about the truth behind Indian culture and how it links science to religion, then I'm sure I'll find them here.

The institute is remarkably intact, like a lot of the architecture in and around this city. The area is overlooked by the lumpy Chamundi Hills, famous for being the site of a mythological battle between an evil demon and the Hindu goddess Chamunda. One legend goes that after defeating him, she made her home on these hills, and now pilgrims must climb a thousand steps if they want to worship her at the golden temple that was built in her honour.

Mysore, at the foot of these hills, is known as the City of Palaces for its grand, pastel-coloured buildings dating back to when kings and queens still ruled these parts. On either side of the wide, sunny European-style boulevards, patrolled by policemen in white cowboy hats, are ornate churches and mosques surrounded by coconut palms and bamboo plants. Standing out like marshmallows in this architectural sweet shop are yolk-yellow and light-green mansions in a mix of styles; some Islamic, some Gothic.

The Oriental Research Institute, although it did not become a government library until 1891, dates back to 1887, when the Maharaja of Mysore commissioned it as a tribute to Queen Victoria, then the Empress of India. It is painted in white and red, with detailed sculptures of Hindu gods and goddesses at the top and two silver lions standing guard at the entrance. Inside, the décor is worthy of an Ivy League campus. The main library has high wooden bookcases, rickety chairs and heavy portraits on the walls. Other rooms are piled floor to ceiling with antique manuscripts, inscribed with tiny characters in Sanskrit.

What hits me first as I enter the institute is an overpowering sickly-sweet smell. The vapours hit the back of my throat and make me gag.

'It's citronella,' says a voice behind me.

I turn to see a thin man crouched over a large wooden desk, eating lunch from a metal tiffin carrier. He was so quiet that I barely noticed him when I arrived, and it's only now that I realise he's been watching me the entire time. His name is Dr Satyanarayana and he is a 53-year-old scholar here at the institute.

'We paint the pages with citronella oil to keep the insects away, and that is what you can smell,' he explains, peering into his tin and scraping out the last grains of rice with a spoon. There is a person in the corner dipping a paddle brush into a tin and sweeping oil over manuscript leaves, one by one. Dr Satyanarayana takes a swig of cold water from a flask and pops it, along with his empty tiffin carrier, into the top drawer of his filing cabinet.

For more than thirty years, funded by the government, Satyanarayana has helped to collect manuscripts that were hidden away in people's homes around India and then brought them here to be studied and preserved. Most of them are still a mystery. In one, scholars found a dried-up white snakeskin, the purpose of which they still haven't been able to explain.

There are around 50,000 manuscripts here now, each made of flattened palm leaves, neatly folded into long two-inch-thick blocks and wrapped tightly in string. This is the second oldest palm-leaf manuscript library in the country. 'In southern India, palm trees are very huge, and it is very easy to scrape the leaf,' Satyanarayana explains. 'They used to scrape the leaf with an iron rod and then they applied a black colour. The palm leaf is not destroyed after many years.'

He raises his head and peers at me over a pair of spectacles perched on the end of his nose. There is a round red *tika* – a Hindu mark of blessing – on his forehead.

'Why are you here?' he asks.

'I'm looking for some works of science.' I'm hoping that among the thousands of manuscripts stored here in the Oriental Research Institute I might find some that explain why those scholars back at the Academy of Sanskrit Research believe there is science in their religious texts.

He nods his head and shuffles out of the door. I follow him into the next room.

'Some are more than 500 years old. And you know there are eighteen volumes of descriptive catalogues. Eighteen volumes!' he says, pointing to a narrow chest of drawers inside the reading room. Inside the drawers are handwritten index cards, each one describing a separate text.

'Are there any on science and technology?' I ask him, flicking through the cards.

'Yes, there are some on astronomy and astrology. And some on gemology' he replies, pulling out a book of English translations for me from a shelf.

'What kind of science do they have in them?' I ask.

'The sun is there, the moon is there, the earth rotate. But at that time our literature was only related to spiritual matters,' he says, wandering off absentmindedly, leaving me to search through the dusty bookcases.

After a while, I dig up some names that I recognise from history books. There's an English translation of the work of Aryabhata, the celebrated Indian mathematician and astronomer who lived around AD 500 and is credited with early work on fractions and trigonometry. And there is a copy of some writings from a few hundred years earlier by the mathematician Bhaskara, explaining concepts from algebra in verse. Flicking through, I find some other books in the science section.

Textbook of Mathematical Astrology, reads one title.

The Indian Ephemeris of Planet Positions.

Span of Life: Astrological Thesis on Longevity. I stop at this one and thumb through the fading pages printed in wonky type. 'If weak moon is in the ascendant which does not fall in Aries or Taurus, aspected by malefics, the child dies very soon,' reads a grim paragraph. 'If the lord of the Rising Sun occupies the 8th, aspected by all malefics, the child lives for four months,' reads another. From what I can tell, the book is about how the myriad positions of the stars and the planets can determine someone's lifespan.

Surfing through page after page, I finally begin to understand what Dr Satyanarayana was trying to tell me. Many of the ideas in these old scientific texts are mixed up with spiritual and religious notions. Genuine theories about how the world works are not separate from murkier pseudoscience. The legitimate search for stronger metal alloys to build sturdier structures is written about in the same breath as the alchemist's pointless quest for a substance that turns base metals into gold, while important ideas about the structure of the solar system are wrapped up in philosophies about how the alignment of celestial bodies can somehow predict the future. Indian astrology, as opposed to modern astronomy, holds that the earth is the centre of the universe.

But then, this shouldn't really surprise me. Many of the old manuscripts in this library were written during the Middle Ages, before the Enlightenment, when science was a very different animal. In medieval times all over the world, scientists weren't even really scientists; they were more like philosophers, picking up ideas almost at random and exploring them independently. This is what makes the work of genuine mathematicians like Aryabhata and Bhaskara so exceptionally important. The experimental method wasn't really invented until about 350 years ago. It took this long for science to develop a recognisable structure, using hypotheses and tests. Before that, it was common for astronomy and astrology to be treated as one and the same.

But the strange thing about some of the books in this library is that even recent ones have the same bizarre mix of science, religion and pseudoscience as the oldest ones. Modern documents dating from as recently as the 1950s and 1960s talk about astrology in the same detailed terms as do the crumbling manuscripts.

After I leave, just down the road from the Oriental Research Institute I spot an advertisement for the services of an astropalmist, who promises to predict people's future by reading the position of the stars and the lines on their hands. Walking into a bookshop in the centre of town, I find heavy titles, published in the last few years, about horoscopes and astrology. There are others on medicinal plant potions for treating diseases like cancer, and row upon row of booklets exploring the links between science and spirituality. Then on television that evening I watch a long infomercial for a set of blue and white beads that promise to ward off the evil eye. I wonder whether the reason for the strange things I had seen at the Academy of Sanskrit Research might be connected to all of this. Perhaps pseudoscience and superstition are as much a feature of everyday, modern Indian life as they used to be hundreds of years ago.

I'm back in New Delhi, this time to meet Sanal Edamaruku, the president of the Indian Rationalist Association. His office is in a residential east Delhi suburb called Mayur Vihar, next door to a shop selling stacks of multicoloured Indian sweets, close to a newly built supersized temple alongside the highway, and a few doors away from a hospital. Bumping through the streets in a green and yellow auto-rickshaw, I'm carrying my notes from Melkote and Mysore.

When I recount my strange experiences to Edamaruku, I expect him to be surprised, but he doesn't even blink. Since he became the general secretary of the Indian Rationalist Association in 1983, Edamaruku has made a career out of debunking pseudoscience, superstition and mysticism. Round-faced with a clipped black goatee, he always manages to turn out a nonchalant smile, no matter how weird the situation he finds himself in. This office is now India's rationalist nerve centre, from where he runs his debunking operation. There's a fat copy of the *Dictionary of World Religions* on the metal shelves behind him, alongside hundreds of other books, some in his mother tongue Malayalam.

Over the years, he tells me, he has collected countless tales of bizarre religious practices and beliefs. Nothing shocks him anymore. Just last night, he says, he was dealing with a case of a guru in the eastern state of Bihar who was reportedly standing on the chests of young infants and reciting mantras, claiming that this ritual would prolong a child's life. Parents were willingly lining up with their young offspring outside his door. When the Association asked local leaders and politicians to intervene and stop the dangerous practice before a child was seriously hurt, none of them wanted to get involved for fear of attracting the wrath of the guru's religious supporters. Edamaruku was forced to turn to news reporters instead. At ten o'clock last night, he says, the guru was arrested.

A fresh case arrived on his desk this morning. 'Now, today in the government assembly of one state, there are people in this assembly saying that there's a ghost,' he tells me.

'A what?' I ask.

'A ghost. A ghost. So they're having an exorcism,' he says matter-of-factly.

Despite being so busy, though, the Indian Rationalist Association has never been a big organisation. There are 207 active groups, Edamaruku says, with almost 100,000 supporters – less than a hundredth of a per cent of the country's population. 'I think everyone is born a rationalist because religion is imposed upon children,' he says. 'And I was lucky that I wasn't having such a situation. My parents were not religious which meant I could have the opportunity to have a liberated childhood.' Born in the famously leftist south Indian state of Kerala, Edamaruku is amongst few Indians to have grown up without religion or any knowledge of his caste. He joined India's rationalist movement when he turned fourteen.

But the Association is becoming popular, he adds. A few decades ago, for example, it had fewer than a thousand members. 'The number of people who come to our lectures or the television programmes where I'm invited to respond is enormous nowadays. I appeared in over 200 television programmes last year,' he says. The phone rings and he pauses our conversation to answer it, arranging an appointment with another news reporter.

Edamaruku's biggest case came in 1995, when a man in New Delhi claimed that a statue of the elephant-headed god Ganesh in his local temple had begun drinking milk. Neighbours rushed to their temples to see whether or not this was true. In the weeks that followed, in shrines all over the world, millions of awestruck Hindus reported that, when they gave an offering of a spoonful of milk to the deity by pouring it near his mouth, the milk would disappear as though Ganesh himself were sipping it. CNN, the *Washington Post* and the BBC were just a few of the media outlets that covered the story in breathless detail. Edamaruku's phone rang off the hook.

When he started visiting temples, however, it gradually became obvious that these stone statues were only sucking up the milk in the same way that anything made of stone would suck up a fluid. Any porous material has tiny holes running over its surface like a sponge, which naturally form thin tubes that draw in liquid by capillary action. A hunk of stone, however, can only take so much milk. The surge of worshippers liberally pouring pints of milk over Hindu statues created a surplus that was running into the drains behind temples. Edamaruku pointed this out. Then he went on national television with a stone bust of Prime Minister Jawaharlal Nehru and poured milk on it. Some of the milk seeped into the bust. Then he poured black coffee over it. And that seeped in too. He repeated the experiment on a brick. The same thing happened. 'The miracle ended with that thing,' he says, laughing.

Actually it didn't. In the summer of 2006, zealous believers repeated the same milk miracle all over again.

'There are two Indias,' says Edamaruku. 'One that is modern, with science, and another India which is living in the Middle Ages. And these two Indias are going at the same time.'

'Do these two worlds ever meet?' I ask him.

'Yes! Some people live in the modern world and periodically they go into the other world. They have two lives. For example, when a scientist consults an astrologer for the appropriate time for the marriage of his daughter, where is the scientific approach there? He's the scientist by profession but he does not use his scientific approach or his scientific mind when it comes to his private life. There are two compartments.'

To understand why this happens, says Edamaruku, I need to recognise how religion works in India. Hinduism – the faith practised by the majority – is 'not seen as a die-hard practising religion,' he explains. Although millions of Hindus may pray regularly at shrines in their homes, and never fail to mark religious festivals, it's a faith that puts few demands on the everyday believer. This separation between the spiritual and the practical is what has allowed people to fit their traditions and religious habits into their normal lives without worrying whether the two conflict. This may even be why India never went through a European-style Enlightenment, which split religion and superstition from science. There simply wasn't a distinction between them.

But modern religious phenomena, like the milk miracle, are different from the way Hinduism used to be practised, he says. Over the years, Edamaruku has noticed that India's biggest faith has become more organised, increasingly resembling more dogmatic religions like Christianity and Islam. And this has all been accompanied by a growing tendency, especially among India's educated middle classes, to try to justify their faith in scientific terms.

Just like the scholars at the Academy of Sanskrit Research who are working to find evidence of science in the Vedas, Edamaruku says there appears to be a growing trend to prove that new ideas or foreign inventions have Hindu roots. He thinks that the reason that religious scholars have come up with stories about ancient flying saucers and alien visits is not just because they're struggling to separate real science from hokum, but because they have an underlying religious agenda to prove that the scriptures are all-knowing.

'I've met a lot of people who simply swallow these kinds of things. They want some kind of scientific justification for their beliefs and if they get a fabricated concept, they will immediately accept it. They want to accept everything that is in modern science and say this was there in our ancient texts,' he says.

He pauses, before adding, 'They want sanction from science.'

This is all very different from the US and Europe, where the opposite seems to have happened. There, devout religious followers have tended to reject scientific ideas that clash with their religious beliefs; in India, some people instead appropriate them. Christians and Muslims, for example, developed the idea of creationism as a kind of countertheory to Charles Darwin's theory of natural selection. They wanted to explain human origin in a way that discounted fossil records and scientific evidence of evolution, which conflict with biblical accounts. And so creationism has become a faith-based alternative to modern science. In India, meanwhile, as science advances, religious scholars are claiming new discoveries for Hinduism by fitting them wholesale into the religion itself – or at least this is Edamaruku's view.

'Earlier they would blindly reject scientific approach of science; now they want scientific proof for what they believe,' he explains. 'Instead of ignoring the theory of evolution, they will say that the seven incarnations of the god Vishnu are like the different stages of evolution. Or for example, nuclear bombs. Many people say that the special arrows, which according to the old legends the epic Hindu hero Arjuna had, they would multiply in power by ten when he drew one. They say that's like an atom bomb or something. People actually defend such positions.'

Edamaruku's observations are supported by Meera Nanda, a historian and researcher who works at the Institute of Advanced Study at Jawaharlal Nehru University on the outskirts of Delhi. She's wearing dark-rimmed glasses and a grey waistcoat when I meet her in her study-cum-bedroom in the mazelike university campus. There are stacks of books against her walls and nothing outside but acres of trees and long grasses. This is almost a hideout. Nanda has spent much of her life studying the complicated relationship between religion and science in India and, in recent years, her work has attracted the wrath of Hindu zealots.

Her argument is that some of India's religious believers have tried to appropriate modern science because they feel uneasy at the thought that there may be phenomena in the universe that are unexplained by their religious texts. 'In this aspect, the idea of Vedic science in Hinduism is actually very similar to creationist science in America,' she says.

In part, she thinks it may have something to do with the rote-learning education system. 'I could tell you some horror stories about Indian education,' she says. 'When I was at school, Indian science was taught in this old scholastic manner, in which there was no critical thinking required. There was no process of forming a hypothesis and doing controlled experimentations. The critical engagement, you know, asking questions, it was just missing. You could go through science education without ever applying the factoid of what you have learned in your science class to the reality of what you are living in. You just don't! They are two different worlds. You just don't relate, let's say, Newtonian law to miracles.'

But the roots of the problem lie even deeper, says Nanda. 'The whole business started in the nineteenth century when we were still under the British,' she explains. 'Indians encountered modern science through colonial education and it was very clear that something quite fundamentally different had emerged in the West which could be empirically tested, and which could be explained without invoking God. I mean you could explain the workings of nature without invoking any creator god, using theories from Newton down to Darwin. So there we were. We were confronted with modern science, we were studying it, we could run those locomotives, we could see its power and we could see those laws worked. We were attracted to it but at the same time, because it was brought to us by the colonial power, there was this aggressive defensiveness about India, about our own culture.'

She goes one step further. 'Indians have a weird psychology, I think,' she adds. 'With respect to the rest of the world, we have an inferiority complex which we hide in a superiority complex.'

This, claims Nanda, leaning forward in her swivel chair, is how Vedic science was born. 'It was a very aggressive way of defending your own faith, you know. And not just defending it, but projecting it as superior.' Spiritual gurus began to argue more and more loudly that the wonders of modern science were something that ancient Hindus already knew. It started with small spiritual ideas, like the quantum physics theory that all the particles in the universe are interconnected, mixed with the idea that human consciousness is also part of an interconnected whole. This is a similar argument to the one used by Vandana Shiva, the anti-GM activist.

Over time, these links between science and Hinduism were built upon to the point of bizarre abstraction. The idea that the gods had floating chariots was invented about a hundred years ago, Nanda says. One religious group 'came up with this notion, a silly notion, that the [holy texts] talk about gods flying in the air. And so came this silly, silly, silly idea that we had aeroplanes, flying technology, centuries ago. It was stuff like that. It's laughable it's so stupid.' Later on, when European physicists identified the atom, Hindu nationalists began to claim that early Indians had already written that the smallest particles were the size of a human hair divided into a hundred parts, with each part divided again into a hundred – coincidentally, not far from the actual size of an atom. Every historical accident and religious metaphor was brought into the mix. Recently, some people have suggested that plastic surgery and biotechnology are Hindu in origin. In some cases, Indian scientists really did make an early contribution to these fields. In other cases, religious believers are simply rewriting history through the prism of their faith.

'Ordinary people could not read their own scriptures,' Nanda adds. So they just believed what they were told. Even now, a new breed of Hindu leaders, known now as modern gurus, is appropriating science and performing seemingly impossible miracles.

Sanal Edamaruku has tracked the practices of these new gurus. 'The traditional Hinduism is dying and a new Hinduism has emerged and the custodians of the new Hinduism are not, for example, the traditional temples or the traditional institutions, but the modern gurus and babas. And they're deciding the mindsets of the people now. This thing happened in the Western world also, if you see the new cults and sects. In the Western world they're simply on one side, they're not on the mainstream. But here, Hinduism has absorbed these people,' he says.

In 2008, after a well-known politician claimed that her opponents had been using black magic against her, Edamaruku attempted to prove that India's new religious miraclemakers were fakes. He challenged Pandit Surinder Sharma, one of the country's most famous Hindu magicians, to kill him live on national television. Bald and dressed in flowing white robes, Sharma had claimed that he could strike any man dead in three minutes just by chanting mantras at him. People were so convinced that he might actually do it that audience figures for India TV, the channel airing the show, soared. He recited one verse after another, each one of which failed. Afterwards, the newspapers printed photographs of Edamaruku laughing. The old magician was standing next to him, furiously chanting in vain.

Even so, Edamaruku concedes that the attraction of miracle men and gurus remains irresistible to many, not just Indians. The popular ones attract more fans than rock stars. The most famous, the late Maharishi Yogi, included the Beatles among his followers. Millions of dollars have also been donated to Mata Amritanandamayi, known as the 'hugging guru' for the supposed healing powers of her hugs. And 83-year-old Sathya Sai Baba claims to have more than a thousand spiritual centres in more than a hundred countries around the world, with millions of devotees. The miracles that he has reportedly performed include holy water and fruit spontaneously appearing out of thin air. In Bengaluru there is even an ashram that holds rock evenings for young IT workers, where they can sing along to devotional music played on an electric guitar and drum kit.

These are more than just religious enterprises; they are multi-billion dollar businesses.

Sunderaj Chandrashekhar, one of the youngest people working at the Academy of Sanskrit Research, is a short, 32-year-old librarian with a long moustache. I've arranged to meet him for a coffee to find out if Nanda and Edamuruku are right. Are scholars at the academy driven by a nationalistic, defensive agenda? Do they think they're right, or are they simply defending their faith against modern science? On the other hand, are their aims even simpler than that? Are they trying to peddle ideas that they know will guarantee them public interest and more money?

'Give the driver your phone, and I'll tell him how to reach this place,' he tells me when I call for directions. The place is a travel agency above a shop, owned by one of his relatives. It's on one of the big shopping streets in Mysore, the city where he lives.

'Please sit,' he says when I arrive, offering me a seat on a threadbare sofa. He spends the first ten minutes trying to get me to book a holiday. 'He can give you cheap tickets.' He gestures to a man sitting at a desk by the window surrounded by curling, blue-tinged posters of beaches and palm trees.

'I don't want a holiday. I'm flying out of Mysore tomorrow anyway. I already have tickets,' I tell him.

'You can cancel them and buy one here instead.'

'I don't want to. Look, if you're not going to help me then forget it.' I pick up my bag and turn for the stairs.

'Sorry, sit down, sit down.'

A woman brings us small cups of tea and a plate of biscuits. 'OK,' he says. 'What is it? What do you want to know?'

'Do you believe that all the information in the world is in the old Hindu manuscripts?' I ask.

'Yes, it is probably. You can create energy for the electricity from the Vedas. That has been proved,' he says.

'Really? How?'

'By the recitational methods. This things. Reciting the Vedas we can create electricity. That has been successfully done. I'll give you all the documents later, I don't have the details now.'

'Anything else?'

'If you go for the test-tube baby, you can refer in the text, where a test-tube baby was born.'

'What does it say, exactly?'

'I can't explain it right now, you have to go and study some of the chapters in the epics. *Mahabharata* and *Ramayana* [other ancient Hindu texts] have them. I don't know exactly which chapters it comes, but one of the brothers was born to a virgin. You have to go see these things. How it took the shape of a man.'

It carries on like this for a while. Every time I ask for hard evidence for one of his claims, Chandrashekhar can't help me. After a while, he looks annoyed and I sense it's time for me to go. Before I do, though, I want to ask him about his views on the *Vaimanika Shastra*.

In 1974, a year after G R Josyer published his famous translation of the *Vaimanika Shastra*, five researchers at the departments of aeronautical and mechanical engineering at the Indian Institute of Science in Bengaluru decided to take a look at Josyer's book. In their paper, entitled 'A Critical Study of the *Vymanika Shastra*', the engineers approached the topic as if it were a piece of genuine science. They took it apart, step by step:

'The height and width of the craft, in our opinion, are in such proportion as to put its stability in serious question,' they wrote. 'It must be pointed out here that the essential idea of flying like a bird has been tried by many people over several centuries right from the time of Leonardo da Vinci, but without any success whatever. Hence the feasibility of a craft of the above type is a near impossibility. Furthermore, the author – whoever he be – shows a complete lack of understanding of the dynamics of flight of heavier than-air-craft . . . Any reader by now would have concluded the obvious – that the planes described above are the best poor concoctions, rather than expressions of something real.'

I wonder if this damning verdict has had any impact on people's perception of the veracity of the *Vaimanika Shastra*. I ask Chandrashekhar if he also believes, like the older scholars at the academy, that ancient Hindus rode flying machines to other planets.

'Aeroplane technology claims to be the invention of the Wrights brothers, from 1965,' he starts (actually the Wright brothers flew their first plane in 1903, but I decide not to correct him). 'When you go back to the history of the ancient technology in ancient India, the root goes back to this one book, written by a saint, where you can see all the details about aeronautical science in ancient India. These are all the texts of the Indian sciences. We call it the Hindu sciences.' He believes it.

'So what do you think about Western science?' I ask.

'That is there. But if you see the books there are references to metallurgy, water management, and astrology, astronomy and other things like electrical energy. All these things, which are referred, go back to the history of the Vedas. Moreover you can see these Vedic mathematics is there, where you can see how things are calculated,' he continues. 'The contribution of the ancient mathematician Brahmagupta's concept of zero. See, without zero you can't even calculate the *pi* of the thing nowadays.'

But there's nothing Hindu about the zero, except that Brahmagupta happened to be a Hindu, just like there's nothing Christian about calculus, except that Isaac Newton happened to be a Christian. It seems to me that Chandrashekhar is trying to drag real science under the banner of his religious beliefs, just as Sanal Edamaruku and Meera Nanda had said.

Before I leave he hands me a science book produced in 2003 by the Academy of Sanskrit Research. It's called *Science & Technology in India through the Ages*. It's hardback, heavy and glossy, with long paragraphs repeating the kind of things he's just told me, except it's liberally scattered with phrases in Sanskrit. 'Looks interesting,' I say, flicking through it. 'Comprehensive.' I hand it back, switch off my voice recorder and get ready to leave.

'You like it?' he asks me, holding up the book.

'I suppose . . .' I'm half out the door.

'Wait!' he calls after me. 'Do you want to buy one?'

On the way from Mysore back to Bengaluru, where the nearest airport is, I stop off at McDonald's for a late lunch. My driver thinks this is decadent, because a burger and fries are so expensive compared to other roadside eateries. And I get the sense he also thinks less of me for choosing an American fast-food joint over an Indian restaurant. Some Indian activists still protest against American chains here for trying to replace indigenous cuisines. But I've been craving a box of chicken nuggets.

Sharing a bench outside with a nonchalant-looking plastic Ronald McDonald, I think about what I've seen in Melkote and Mysore. There are lots of reasons why people peddle new ideas or inventive products. Sometimes it's raw commercialism (like McDonald's), while other times it's nationalism or pride (like the people campaigning against Big Macs). But, when it comes to religion, neither of these reasons feel as if they reach the heart of the issue. If commercialism and nationalism alone could account for the rising popularity of miracle gurus and magic charms, that would mean that every one of India's billion believers were just suckers for a good sales pitch.

There must be more to it. I get the sense that part of the reason lies in *who* these believers are. It stands to reason that people in rural villages, where rates of illiteracy are high and lifestyles haven't changed for centuries, might believe in religious myths and superstitions. But Nanda told me that, unlike in Europe, where rates of religious observance have steadily dropped over the years as literacy and education improves, India is actually becoming more religious. And she claims that it is in the modern cities, not the rural areas, where rates of religious observance are rising fastest. Clever, wealthy people are the ones paying millions to miracle gurus.

'I think what's happening is that as Indians are getting more educated, they're getting more sort of *scientistic*,' she had explained to me before I left her office. 'For an ordinary believer, it's just faith. They don't need to explain it. But there's a certain class of people coming up that need to justify their faith, who need to somehow intellectually put into words why they believe. It's more of a disease of educated people.'

But surely education makes people smarter, not more gullible? And this makes me wonder. Maybe Nanda is right, or maybe the opposite is true. Could a portion of this well-informed generation of Indians be attempting to make their beliefs sound more logical and scientific, not because they believe them more, but because they fear that they believe them less? Perhaps it's an attempt to rationalise what, on the surface at least, can seem irrational.

I have one last meeting before my flight. It's with Professor Mathur Ramabhadra Shastry Narasimha Murthy, the head of the molecular biophysics unit at the highly respected Indian Institute of Science in Bengaluru. He is well known in Karnataka, where can be found both the institute where he works and the Academy of Sanskrit Research, for countering popular superstitions with good science.

His institute has a huge campus. It's approaching the end of the day and the trees are casting long shadows over the wide walkways. Activity in his small laboratory is winding down. One girl at a workbench pipettes a liquid into small round containers, before changing out of her white coat and leaving for home.

Murthy offers me a seat next to his laptop. He is a softly spoken, disarmingly charming man, wearing a stripy shirt with a pen tucked into the pocket, and round glasses. Behind the thick lenses are happy, baggy eyes. They're the signs of a man who, although satisfied with his life, I suspect has had trouble sleeping at night. When he was younger, he tells me, he struggled with his faith when he started to study modern science. He grew up in a traditional and devout Hindu family in rural Karnataka, but came to the realisation that the rational world had no place for the supernatural. He ditched religion and became an atheist.

'What's it like being an atheist in India?' I ask.

'To be a true atheist is a deadly position. You have to redefine your whole life and its meaning. I am in a minority. A small minority, whether in the East or the West. I have come across many Europeans who claim to be atheists but they have some element of belief. I had a particular colleague in America, and he refused to take a job because the last guy, when he did this job, he developed a leg problem and couldn't walk. It was bad luck, he said. I took the job and nothing happened to me!'

There is something about superstition, he says, which people across the world seem unable to abandon. He was reminded of this just a few weeks ago, on the fifteenth of January. It was the day of the longest annular solar eclipse of the millennium and south India had an ideal view. This special celestial moment comes about only when the moon, sun and earth line up perfectly, but the moon's shadow is slightly smaller than the circle of light from the sun, giving the effect of a glowing ring of fire in the sky. For most Indians, it was a must-watch moment.

But others were cowering in their homes, afraid that the eclipse might harm them. Some astrologers had warned that the eclipse was dangerous to foetuses, so pregnant women reportedly stayed indoors. The astrologers also said it was an unlucky day to cook, so thousands of people fasted all day.

Murthy decided to organise a telescope meeting to encourage people to watch the eclipse. 'The eclipse has been occurring since the solar system was formed,' he told newspaper reporters, attempting to reassure the masses. 'No harmful radiation occurs during the eclipse. There's no reason to refrain from doing anything, even eating.'

In his heart, though, he must have known that this wouldn't change people's minds. Belief in astrology isn't just something that happens on the fringes of Indian society. Eight years ago, for example, the Indian government – then controlled by the Hindu nationalist Bharatiya Janata Party – tried to introduce astrology as a new subject in university science departments across the country. It was an attempt to insert traditional Indian philosophy into modern scientific teaching. Professors and researchers protested, so the plans were shelved. Then just a few weeks ago, the local government here in Bengaluru attempted the same thing again. Heads of university departments were again forced to complain, describing the idea of having undergraduate courses in astrology as an 'obscurantist and unscientific move.'

I ask Murthy whether he has heard of the *Vaimanika Shastra* or read the Hindu scriptures, like the Vedas.

'Actually I know the Vedas quite well,' he says.

This surprises me. It turns out that he's not just an atheist; he's an informed atheist. He has spent hours of his spare time studying the sacred Vedas. Learning Sanskrit alone must have taken him years.

'You know we have to be careful when we say that there are elements of scientific thought in an ancient work,' he says. 'Certainly you do not have ideas of particle physics or aeronautics in the Vedas. This is junk if somebody says that. First of all the meaning is not a hundred per cent clear, because it is in an archaic language and we do not fully understand it. It's not modern Sanskrit, with which we are very familiar.'

'Do you think there is any science in the Vedas?' I ask him. I'm expecting a 'no'. He surprises me again.

'Depends on what you mean by science. The Vedas certainly speak of astronomy. And I tell you, there is very beautiful verse in the Vedas, which speculates on the origin of the universe. Maybe the energy was condensed and maybe from this energy the whole world arose. Maybe it did not. Who can tell us? Maybe the gods will tell us, maybe we will learn. It is beautiful poetry.'

But then his tone changes. 'You can show or you can think there are parallels to Big Bang Theory in this, but certainly there is nothing like Big Bang in there. It is nothing like Big Bang Theory.'

'Then what is it?'

'It's just the human imagination was extremely powerful. You take any modern scientific idea and you can see in the midst of early culture, elements of thought which are resembling modern science. But science was not there. See, it doesn't take much to imagine, looking at birds, that some day we might learn to fly. But if somebody claims that there are real flying machines in the Vedas, well it's just stupid.'

A boy comes in to deliver him a letter, interrupting his train of thought. 'Look, what I call modern science is what happened after Descartes and Newton,' he continues after a pause.

'After the Enlightenment?'

'Yes. That's what I call modern science, Murthy says. So if you go to these old Hindu astronomers, they will start their treatises on astronomy with a prayer. This is not done in modern science. You read a textbook today, it doesn't start with a prayer to the Lord to save humanity and then start the science. Modern science strictly banished God from science. We removed any supernatural explanation from all our explanations. We had to account for laws of nature by what is observable. Scientists working in India now are working in the European tradition of science, in the British and French tradition of science. So in some sense we have adopted science. It's not homegrown, it's adopted.'

'But not everyone's happy with adopting it,' I interrupt, thinking back to what Nanda had told me about the defensiveness of Indians who felt uncomfortable with scientific ideas that came from former colonial powers.

'Yes. But an adopted child can be as dear as your own child.'

The fact is, though, that India does still live in the shadow of colonialism, and the lingering awareness of racism and sense of inferiority that comes with it. As Nanda suggested, it seems a likely explanation for the country's ongoing tussle between the modern and old, between science and religion, and logic and superstition. It also seems to be at least one of the reasons behind the strange ideas I encountered at the Academy of Sanskrit Research and the rise in religious observance in urban areas.

Behind Murthy's happy, baggy eyes, I can see how hard the tension between the two must have been for him. I had thought that he had blindly chosen science over religion, assuming that a modern, atheistic scientist would never have found value in something as archaic and mystical as Hindu religious texts. So it's both humbling and reassuringly geeky to find that he didn't dismiss out of hand the religion he grew up with, but instead studied it in detail and drew his own conclusions. Murthy sees the scriptures objectively for what they are. And to him they are no more than beautiful poems that explore both the living reality and the imaginary, asking questions rather than offering answers. When sixty years ago in the Indian constitution, Prime Minister Jawaharlal Nehru called for people to 'develop the scientific temper', I imagine this is the kind of logic he wanted them to apply to their everyday lives.

Murthy looks at his watch. 'I need to leave now. I'm babysitting my granddaughter,' he says, glancing proudly at the picture of her on his computer's screensaver. I offer to walk out to his car with him.

It's unfair, he continues, of outsiders to think of India as an intrinsically superstitious or irrational country. 'You know, let me tell you, humans everywhere in the world irrespective of race and religion are equally sensible, equally foolish. It's like good people and bad people,' he says, flashing me a broad, toothy grin. The major reason he offers for people clinging to their superstitions is illiteracy. 'I have no doubt in my mind. General, secular education is a very good cure for superstition, and of course many Indians are illiterate. So if they are educated, superstition will be less. In most countries in Europe, literacy is a hundred per cent. So to that extent if you take into account those who are educated in India, the level of stupidity in them is very comparable to Europe. I think Indians on average are as sensible as people anywhere else,' he adds, stuffing some papers under one arm and locking the laboratory door.

History has been full of instances where people all over the world have been unable to distinguish real science from pseudoscience, he explains. Back in 1895, Lord Kelvin, the President of the august Royal Society of Great Britain, said that heavier-than-air flying machines were impossible. Within a decade he was proved wrong. People are continually duped by ideas that sound solid on the surface but later turn out to be hokum.

Among Murthy's favourite examples is the nineteenth-century art of phrenology, which claimed that it was possible to read someone's personality by feeling the bumps on their skull. Phrenologists thought that different parts of the brain were bulgier depending on how active they were, so a person's head might be bigger in one place if they were arrogant, say, or in another place if they were particularly poetic. They thought that the brain had around thirty distinct organs in all, each linked to different faculties, from digestion to memory.

Murthy climbs into his beaten-up little red car. 'You know, I think the world is quite irrational,' he says, starting the ignition. 'It is not to do with India, it is to do with the human race.'

CHAPTER 5

THE MINDREADING MACHINE

For anyone who wants proof of how irrational even educated Americans can be, I know a good story. It starts in a courtroom in Washington DC on 3 December 1923, at a murder trial.

A rich doctor had been shot dead and a few months after his killing, the police picked up a notorious armed robber named James Alphonso Frye, who became the prime suspect. It looked like an open-and-shut case, and it would have been, except that Frye claimed to have scientific proof of his innocence. Dr William Marston, a Harvard-educated psychologist, had invented one of the world's first lie detectors (then known as a systolic blood pressure deception test) and Frye was his guinea pig.

The procedure was simple. Frye was strapped into a blood-pressure cuff that measured the subtle changes in his blood flow while he was being questioned.

But sadly for him, passing the test wasn't enough to spare him from jail because the judge wasn't convinced by this mysterious new invention. Frye was given a life sentence for murder. He died thirty years later. Meanwhile Marston abandoned his lie detector and instead turned his hand to writing comic books. Fittingly, he became the creator of Wonder Woman, a superhero who whips honest responses out of her enemies using a Golden Lasso of Truth.

This could be the end of the story, but it isn't. Just before Frye died, the US became obsessed with lie detection. This was a new age of space travel and atomic energy, when technology was believed to be infallible, and so scientists invented the polygraph. It was a contraption that measured a host of 'signs' that indicated whether a person was lying, including pulse rate, breathing rhythm and skin conductivity, as well as blood pressure. American intelligence agents, police officers and employers all began using it. By the 1950s, around two million polygraph tests were being carried out in the US every year. The public believed that it was finally possible to read the mind.

The problem was that, just as with Marston's original machine, the polygraph wasn't as accurate as people assumed. Scientific experiments revealed that it had a success rate of less than 80 per cent, and that was just among ordinary people, not hardened criminals or psychopaths who might be able to cheat it. It fell out of favour. These days it's unusual to see polygraph tests anywhere other than on daytime talk shows like *Ricki Lake*, and in some parts of the US government. Scientific scepticism has almost killed the polygraph, as it did the mysterious practice of phrenology, which was so popular in Europe in the nineteenth century. It's a tale that serves as a warning to anyone daring to believe that they can read the human mind.

But then, they do say that history is cyclical.

It was Europe's turn in the nineteenth century, the turn of the US in the twentieth, and now India has been bitten by the belief that mindreading is possible.

Before I arrived, I'd read a few news reports about a new kind of lie detector being used in a high-profile and controversial Indian court case. The public is still gripped by this story of an atypical murder suspect – a quiet young woman, Aditi Sharma, from a middle-class family – who was hooked up to a truth machine in a government forensic laboratory in Mumbai.

The last thing I heard, she failed the test, and now it's being used as evidence against her.

Outsiders often still think of Mumbai, the biggest city in the western state of Maharashtra by its former name, Bombay, or as Bollywood, the beating heart of India's booming movie industry. Driving down the streets in a crumpled black Ambassador taxi, there's something theatrical about the sweeping view of the ocean from the city's promenades and its motheaten collection of art deco apartment buildings. The people look a little more stylish and sound a little louder than they do in other Indian cities, as though they're all living through an audition for a film part.

But behind the dramatic backdrop, the reality is seedier. This is a city synonymous with slums, gangsters and terrorism.

When I arrive, it has been only a short while since Mumbai was sent into panic by Islamist terrorists. They targeted the iconic Taj Mahal Palace Hotel, only metres away from the Gateway of India on the grand seafront, from which old-fashioned horse-drawn carriages ferry tourists around. The tragic attacks left 167 people dead, and forced the government to question how safe the country really was. One problem is that India has one of the most overstretched police forces in the world. According to the most recent figures from the United Nations, there are only 1.22 police officers to every thousand Indians (the average in Asia is 1.35, and in the United States it is slightly more than 2.25).

So over the past decade, the government has trained up a cadre of forensic scientists to help the police with their bulging caseloads. The hope is that they can speed up investigations by pinning criminals to their crimes using the latest technology.

And that's why here, in the Directorate of Forensic Science Laboratory in a slummy district of Mumbai called Kalina, scientists are experimenting with a truth machine.

The directorate is a large blue office block, built on a drained mangrove swamp that has turned the air into a thick, humid breeding ground for fat mosquitoes. One zips straight for my face, landing a sucker on my eyelid, which promptly swells up into a red ball. A moustachioed, khaki-suited police officer (this place is crawling with them, rifles slung over their shoulders, like extras in a Bond film) laughs when I lift a tube of insect-repellent spray out of my bag.

'Girl, that won't work here! These are Mumbai mosquitoes. They're big and strong, like the criminals,' he jokes.

'I have to try though don't I?' I start spraying.

'Why are you here?'

It has taken me weeks to get permission to come here. 'I have a meeting with Dr Sunny Joseph.'

His directions to Joseph's office are complicated, taking me through a winding set of corridors, upstairs, and through some more corridors. I pass busy toxicology and ballistics labs, which are sending wafts of pungent chemicals (chlorine, perhaps, or ammonia) through the building.

On the same floor as the truth machine there's a carpenter, scattering sawdust into the stairwell. The building feels like a work in progress. Although it's routine for forensic teams in Western countries to collect fingerprints, DNA, sperm and blood samples from crime scenes, forensic science is fairly new to India. Back in 1897, the Kolkata police force did set up the world's first fingerprinting bureau – before even Scotland Yard in London – but for some unknown reason, this early foray into forensic science didn't spread to the rest of the nation. When a terrorist bomb hit a market in New Delhi in 2008, for example, a terrorist bomb hit a market in New Delhi in 2008, for example, newspapers reported that officers rushed to clean the crime scene with buckets of water, which meant that, by the time investigators arrived to collect the forensic data they needed, there was none left. In fact, many Indian police teams don't use forensics at all.

This makes Maharashtra one of the few Indian states to have invested in cutting-edge equipment and qualified staff. The Kalina forensic-science laboratory is a model for the rest of the country.

'We do possibly fifteen cases a month. Ten cases definitely. Referral rates have been really high these days,' says Sunny Joseph, the assistant chemical analyser, when I arrive at his small, stuffy office. He has a neatly clipped moustache and a perfectly pressed shirt. Although his job title makes him sound as though he analyses chemicals, he actually works as a psychologist and is one of the operators of the truth machine. This department has two other staff members, both female psychologists in their twenties, who are rushed off their feet. Case reports in brown folders, indicative of the backlog sit in heaps in the corner of the office. And builders outside are sawing through sheets of medium-density fibreboard in busy preparation for a second truth machine, which is waiting in a spare room, covered in bubble wrap.

They're proud of their results. 'We have never had a case in which there was contradictory evidence, where police had something else to say, or there was some physical evidence contradicting what our findings were,' Joseph tells me.

Before joining the Directorate of Forensic Science here in Kalina, Joseph worked in a hospital with patients suffering from obsessive compulsive disorder. It was a good training ground for the work he does now. 'This is a disorder in which people keep getting repeated thoughts and tension builds up. To break down the tension they perform certain actions repeatedly,' he says. They might keep washing their hands, or checking that a door is locked, for example. For some of these patients, the reason for their symptoms lay in abnormalities in a particular corner of the brain near the back of the skull, known as the anterior cingulate. And this taught Joseph the basic principle behind the truth machine used in this laboratory: that many aspects of human behaviour, personality and action are localised in sections of the brain. Not quite in the order that phrenologists imagined centuries ago, but actually still quite specific.

Obsessive compulsive disorder is just one example out of hundreds. The celebrated American neuroscientist Vilayanur Ramachandran (who was born in Tamil Nadu, in India), for example, once studied a patient who had suffered a stroke in the right half of her brain. Afterwards, her family was shocked to find that she would comb her hair, paint her nails and apply her makeup the same as before the stroke, but only on the right-hand side of her body. She completely ignored the left half of her world. Even when she was asked to draw a flower, she sketched all the petals on the right side of the stalk but none on the left. The reason for the problem, Ramachandran realised, was that the stroke had damaged her right parietal lobe, which is the part of the brain near the top of the head that allows people to judge the spatial layout of their environment. Once this corner of her brain was broken, her mind blocked out everything to her left. To an outsider, her behaviour seemed crazy, but in fact it was evidence of how contained different functions in the brain can be.

Neuroscientists, for example, now know that the lateral hypothalamus (near the front of the head) is the small section of the brain that is sensitive to glucose levels and therefore controls how hungry we feel; that the posterior nucleus of the hypothalamus controls our response to cold temperatures to such an extent that, when it is severely damaged, patients suffer hypothermia; and that the hippocampus (in the centre of the brain) is linked to how new memories are formed and older ones are retrieved. Ramachandran has even pinned down some people's claims of religious experiences to seizures in the temporal lobe.

Detailed brain research like this was exploited to build the truth machine being used here in Kalina. But it is more complicated than that, Joseph tells me. It is all about the memory of the crime. Thousands of papers have been written about the particular parts of the brain that are linked to recalling memories. He reels off a few for me, cupping different parts of his head as he goes: 'There's the prefrontal cortex' – he holds his forehead, 'the anterior cingulate' – he points through the front of his head, 'and related limbic structures' – he moves his hand around from the front to the back of his skull.

I interrupt. 'So how does the machine itself work exactly?'

'Wait a second.' Joseph's office is divided into two sections so that he and his colleagues can have privacy if they need to go through confidential police files or interview suspects. He disappears into this second section to make a phone call, coming out after a few minutes with an offer: 'Do you want to see it?'

'Yes, please.'

He takes me to the neighbouring laboratory, unlocking the door. A blast of dry, cold air hits us. There's an air conditioner inside that ensures that the laboratory remains a few degrees below room temperature, to stop the expensive equipment from overheating. The ceilings are low, there are no windows and the walls are painted white like a dentist's surgery. At the same time, perhaps because it is so small and cold, it reminds me of a crypt.

The laboratory, like his office, is divided into two sections, but this time separated by one-way glass. In the first room is a large brown chair, facing a television monitor, with a blue metal box behind it covered in flashing lights. There is one cupboard with its door slightly ajar, so I take a peek inside. On a lonely shelf is a ball of cotton wool and a syringe. In the second part of the room, where the operator sits, is a giant hard drive, almost as tall as a twelve-year-old, with a set of swivel chairs and matching black desktop computers.

Each one has the same screensaver: a green neon fingerprint and the acronym BEOS, which stands for Brain Electrical Oscillations Signature test.

In this very room in July 2007, Sunny Joseph's team ran the test that I had read about in the newspapers. The suspect was a 24-year-old student from a respectable family, Aditi Sharma, and her alleged victim was her ex-boyfriend Udit Bharati. She had poisoned him, the prosecutors claimed, by taking him to a McDonald's restaurant and giving him some sweets laced with arsenic. Her motive, they said, was that she had gone to college and fallen in love with another man. Sharma was supposed to be getting married to Bharati and he became wildly jealous. So, allegedly, she decided to get rid of him.

Sharma had been languishing in a high-security jail for so many months before leaving for her lie detector test that by the time she arrived here, she barely spoke. She was given an oversized white apron to wear (since most of the people who wear it are older men, it was a little too big for her small frame) and asked to sit in the chair, in the room next door to where Joseph and I are now sitting. A technician squeezed sticky conductive paste through a blunt syringe into a small button on a red skullcap that sat tightly on her head. And this was repeated thirty times until every thin wire sticking out of the cap had made an electrical connection between her brain and the blue box on the table behind her. She sat for more than an hour, watched through the one-way glass.

And then the test started.

For forty-five minutes the psychologists sitting next door played a series of prerecorded statements in Hindi, in an attempt to find out if Sharma really had the mind of a killer. She wasn't supposed to speak, just listen to each statement and let her brain silently do the talking. For a while, nothing happened.

But then there came a statement: 'I got arsenic from the shop.' Suddenly the device registered a response.

'I called Udit.'

It lit up again.

'I gave him the sweets mixed with arsenic.'

And again.

'The sweets killed Udit.'

And again.

Meanwhile the Brain Electrical Oscillations Signature equipment measured the frequencies of the electrical signals from the surface of her scalp and fed them through a set of rainbow-coloured wires into the room next door, where a processor performed a set of complex calculations. The printer spat out its conclusion in red letters. Aditi Sharma was guilty.

The following summer, when her case finally came up in court, she was sentenced to a lifetime in jail. She didn't realise it at the time but she was the first person in the world to be convicted of murder based on evidence that included a brain-scanning truth machine. In the six months that followed, two more people were convicted and sentenced to life imprisonment by evidence that included tests from this same lab in Kalina. One was a shop worker who had allegedly hammered a colleague to death. The other was the notorious 'beer man', a suspected serial killer who reportedly left empty beer cans beside the bodies of each of his seven victims.

'Would you like to try it out?' Sunny Joseph asks me, pointing to the large wooden seat where Sharma had sat.

It reminds me of a replica of an electric chair I once saw at the London Dungeon. The museum staff had strapped a ghostly pale mannequin into it, as if it were ready for execution, and then lit the chair from underneath to multiply the horror. 'No, thanks,' I say.

'Sure? I've tried it. It's nothing!'

'Does this machine really work?'

'Yes. We consider the brain as a computer, where information is stored that can be retrieved,' he says.

I ask him whether suspects are scared of sitting in that bare white room, having their fate decided by a machine, however powerful it may be.

He pauses, swivels away from his computer monitor and looks squarely into my eyes. 'I'll tell you. I'll tell you. They are so, so much relieved to be here. They're so much happy to be here with us, because we're not scary. We talk to them nicely. Just imagine,' he says. 'This is better than the police. You can imagine in India the way the police must be dealing with them.'

It takes a machine to read a machine. And the hardware behind the Brain Electrical Oscillations Signature test is a common, hundred-year-old piece of technology known as the electro-encephalogram. It comprises a skullcap and wires, and it detects tiny fluctuations in brain activity. It works because the brain sends tiny but measurable nerve signals to different parts of the body using small electrical impulses (each between a nanovolt and 100 microvolts, which is roughly the amount of power needed to light a hundred-thousandth of a light bulb). From the scalp, it's possible to detect these tiny signals. The drawback is that the readings are fuzzy, making it difficult to pin down activity to an exact part of the brain.

Nevertheless, neuroscientists often use an electroence-phalogram to find out which parts of the brain are more active than others and why. Experiments have shown over the years that, for example, different mental states produce their own frequencies. Low-frequency brainwaves are associated with a drowsy state of mind, while a brain that's alert pumps out higher-frequency waves. Joseph shows me some coloured cross-sections of what the brain looks like during an electroencephalogram.

There are red blobs of high activity scattered over the picture, against a green background of low activity.

This is not even the first time that an electroencephalogram has been used for lie detection. Twenty years ago, a Seattle-based American scientist called Lawrence Farwell began to research the links between deception and an electrical brain phenomenon known as the P300 wave. The P300 is a positive voltage bump on an electroencephalogram reading in the parietal region of the brain, near the crown of the head, which appears a fraction of a second after a person sees something new or interesting. For example, if someone is shown a picture of a square for a few minutes, the electroencephalogram readout won't change after a while because the picture becomes old and familiar. But if the square suddenly gets one rounded corner, that's when you're likely to get a P300 response. Farwell guessed that this could be used in criminal investigations. The police could show a suspect some photographs of different crime scenes, and when he or she spotted one they recognised or which had some personal significance to them, their brain might produce the characteristic P300 bump on an electroencephalogram.

Farwell operates a business in the US that offers tests using his approach, but he's failed to get his method accepted as evidence in America's highest courts. The technology, critics say, is simply not reliable enough yet to decide whether or not a specific person is telling the truth in every case.

In India, the story is different. The inventor of the Brain Electrical Oscillations Signature test – a psychologist from Bengaluru called Champadi Raman Mukundan – used research similar to Farwell's when he was developing his software. Joseph tells me that Mukundan created a complicated set of algorithms, which process the data that comes out of an electroencephalogram (not just the P300 wave, but other factors too), and that he uses this to decide whether or not the suspect is telling the truth.

'There is so much to it,' says Joseph.

The way it works, he explains, is by linking the suspect to a specific experience and his or her memory of it. 'The system is not just telling us this person has done this murder. It has to tell us so many other things. Like, what is the extent of his participation? And was it only taking the knife and giving it to another person? Things like this. Now suppose there are ten parts of the brain activated when an experience is processed. An experience includes so many components. Experience includes all my visual imageries, to remember this particular event after seven days, after ten days. And then auditory images. And also all motor imageries, meaning sometimes the suspect might have been stretching his hand, like to use a knife or a gun.'

If Aditi Sharma hadn't carried out the crime herself, the test would come up negative, Joseph continues. Experiences can't be planted in the mind by police officers or lawyers, for example, because the stories and experiences are stored in the brain in different ways. A story is just a set of words, whereas an actual event involves actions, smells and sounds. 'Now we are sitting talking here. This is an experience for me. This is an experience for you,' he says, pointing to his chair and then my chair. 'Now you go back home and tell your friend. Whatever we discuss here, whether it is four or five hours, you can only impart knowledge of your experience of sitting here. Your friend can never have the experience of sitting here unless she comes and sits here. This is how it works. This information will be stored in the brain only if we undergo an experience.'

But I'm sceptical. Not only is every human brain different, but criminals in particular are more likely to have aberrant mental states. Psychopaths and pathological killers, for example, often show signs of brain damage. Memories also change and fade over time.

But Joseph has no doubts. He begins clicking through some old case notes on the computer next to him. People know this system is so good that 'we have had at least ten cases out of the

75 or 80 Brain Electrical Oscillations Signature tests we have had in which we have had confession,' he says.

'So you're pretty confident about its accuracy then?'

'Yes, completely. Compared to the other case reports, Aditi Sharma's experiential knowledge was very less. But whatever we had was very significant,' says Joseph.

'Were many of the others accused of murder, like Aditi Sharma?' I ask.

'Yes, most of them murder, yes.'

I ask him to tell me more about Champadi Raman Mukundan, the inventor of the Brain Electrical Oscillations Signature software. Joseph's face lights up in admiration. 'He knows so much about the brain!' he says. 'You know, he even made his own electroencephalograph! Back then, there was no equipment, no funding. This was about thirty to forty years back. He did it because he was very good at physics and electronics and things like that. So with the help of some people he made his own electroencephalograph.'

I meet Mukundan at the forensic science laboratory in Gandhinagar in the neighbouring state of Gujarat, where he works occasionally as a consultant.

Like Mumbai, Gujarat is a scientifically forward-thinking place. Just recently, state officials here opened a Science City theme park on an empty plot of land in the nearby city of Ahmedabad. Unlike Mumbai, however, it's also deeply conservative, and is governed by a chief minister infamous for his radical Hindu views. Drinking alcohol is forbidden. Cow-killing is also heavily policed. One of the pieces of kit used by the forensics department here, for example, is India's first mobile meat-testing facility, which police crews use to hunt down clandestine beef butchers.

The laboratories here, like the ones in Mumbai, are fond of using experimental techniques. Their Brain Electrical Oscillations Signature facilities are even bigger than the ones in Kalina. They've been using them since 2003. In the first five years, they tested 163 subjects in eighty-eight criminal cases.

A woman working at the Gandhinagar Forensic Laboratory tells me that, pretty soon, this city will also be home to the world's first university dedicated to forensic science and criminology. 'There will be everything there,' she says enthusiastically, handing me a leaflet about the university. 'Just twenty kilometres from the international airport.' I flick through it: 'Ballistics, DNA finger printing, toxicology, psychology, blood alcohol analysis, document examination, cyber crime lab, photography, narcotics, explosives, latent fingerprint, wildlife forensics . . . four cow-meat testing mobile vans . . .' it reads. And of course students will also be trained to use the Brain Electrical Oscillations Signature equipment, she adds, showing me to the laboratory where Mukundan is sitting.

He is a shy 67-year-old with salt-and-pepper hair. Immediately, he reminds me of a kind uncle and I can't help but like him.

Through the one-way glass next door is a big chair that has three skullcaps hanging on the wall behind it, looking like dead squid. I feel uncomfortable, just as I did in Mumbai. I imagine the fear of the person who has to put one of these skullcaps on, feel the cool paste from the syringe meeting their warm scalp and then the sticky goo left in their hair when the cap is finally peeled off. I'm surprised that being in this room doesn't seem to bother Mukundan at all.

In fact, he's chirpy. 'In my family, they say that I was this boy born with a screwdriver in his mouth,' he says, smiling. 'From my childhood I was a mechanic, and I was a dangerous type of mechanic . . . I used to get punished every day. If I find I need a plank of wood or something, I go and I remove the plank from the back of the wooden wardrobe in my house. I remove it and take it. Suddenly my parents open the wardrobe and find there is no . . .' he dissolves into a belly laugh. He tells me this is the reason he ended up studying physics and mathematics at university.

Later, though, he dropped physics for psychology. 'It was only when I did my masters degree that I realised I have done a mistake. I didn't know what to do with myself. That's when I joined this. But then nobody understood why I took to psychology. That seemed to be against my spirit.' He was, by his own admission, the weakest student in his year. Unlike the others in his postgraduate clinical psychology class, Mukundan avoided meeting patients. He claims to have seen fewer than seven in his entire degree. 'In clinical psychology you have to spend a lot of time with other people, sitting with them, talking to them. After joining, I found I'm not interested in that,' he admits.

He preferred the laboratory, where he could tinker with electrical equipment. For his PhD project, he hooked up patients suffering from schizophrenia to a galvanometer that measured the electrical resistance across their skin when he applied a small current. Resistance is an indicator of emotional stress, he explains.

Five years later, he singlehandedly founded India's first electrophysiology laboratory. 'The body produces electrical activity,' he explains. 'Measuring that electrical activity associated with the physiological changes is what we call electrophysiology. When there is a physiological change, there's also an electrical change, for example heart rate.' He patched together old bits of kit for five years until he understood it. He says he was a maverick who fascinated the students and puzzled his colleagues.

For years, Mukundan pored over academic papers about how the brain processes memories. And then he started work on the Brain Electrical Oscillations Signature test. 'See I am basically an electronics man, not a psychology person. So it was easier for me to conceptualise the neural organisation,' he says. He claims to see things that other psychologists and neuroscientists can't, because he is a 'hardcore reductionist', he says, paring down the human mind to its nuts and bolts like the machines in his laboratory.

'Is that why you think reading the mind is possible?' I ask.

'If you allow me to elaborate a little, I would say that one day we will have an explanation for everything. Knowing means it's an explanation using certain concepts. So we may have an explanation for everything. And we may even be able to control many things.' He pauses. 'That's more important than just knowing, you know.'

'So you don't believe that the mind is more mysterious and unknowable than that? Do you believe in a soul . . .' I begin to ask.

He cuts me off before I finish the question: 'I don't believe in a soul.' He's an atheist, he says. 'There's this famous biblical saying that God created man after his image, in the image of God. And man later said that he created the computer after his own image – the brain, I mean. I would say that man created both God and computer.'

I still find it difficult to believe that Mukundan can so easily strip down the brain and so simply extract a person's memories, as if the mind were nothing more than a hard drive. But he's a convincing man. Nine years ago, when no other company was interested, he persuaded a team of twenty-something computer engineers in Bengaluru to design the algorithms that would underpin the Brain Electrical Oscillations Signature software. They worked for four years without pay, funding themselves with money they earned from other programming projects. One of the young programmers even lived with him and his wife for two years. Whenever Mukundan woke up during the night, plagued by some problem with the software, he'd wake up the guy too and they would spend hours working through it. The young team collected all the data Mukundan had found on the links between the brain and memory, and translated it into a set of eleven physiological variables. If all eleven variables come up positive on an electroencephalogram, then the statement being read out to a suspect was assumed to be true.

'How exactly do these variables work?' I ask him. 'What are they?'

'The software knows where are the things, where it has to go to look. It's complicated,' he says.

This vagueness is partly why scientists in India have raised concerns about Mukundan's truth machine. In September 2008, a report by a committee at India's National Institute of Mental Health and Neuro Sciences, where Mukundan used to work, announced that brain scans of criminal suspects were unscientific. It warned that they should not be used as evidence in a court of law. The director of the government forensic science laboratory in Bengaluru, Dr B M Mohan, has also opposed the use of Mukundan's truth test. In a phone call, he told me, 'If you ask them which oscillation they're using, they do not tell you. It could be a sea of oscillation.'

Clinical psychologist Dr Shripati Upadhyaya, who was Mukundan's classmate at university and now works for the National Health Service in England, is sceptical too. 'There are some people who are out there to make money from this technology,' she says. 'Experiential knowledge is in the realm of psychology. You can't deny it, unless someone has had a brain injury or taken mind-altering drugs. But to understand the difference between a positive signal and a negative one you need some control data. And I just don't believe that there is enough control data out there yet.'

But these concerns haven't affected Mukundan's success. After it was developed, five Brain Electrical Oscillations Signature kits were sold to forensic labs across India. Another one is on its way to the northern city of Chandigarh. The glossy brochure advertising the equipment has a picture of a face on it, with a magnifying glass above the skull and thin wispy lines drifting out of the brain towards the magnifying glass. 'Get a suspect's brain to talk to you,' it says on the first page.

'They think they can sit and fool us,' Mukundan laughs, looking through the one-way glass. Once, he says, a man arrived for his test having drunk an entire bottle of cough syrup in the hope that this might skew the electroencephalogram readings. He had to retake the test on a different day.

Another time, a suspect sat in the chair and prayed.

Back in Mumbai, I take a taxi to meet the lawyer representing Aditi Sharma, the student accused of murdering her ex-boyfriend. Revati Dere has cropped hair and dark kohl around her eyes, and lives in a small apartment in the west of the city.

It's been a long time since she looked at Sharma's case notes, she says, sorting through the papers on her coffee table.

'I find it very strange that something can read the mind . . . it's outrageous really,' she mutters. 'Somewhere, someone down the line should understand it's the human mind that you're talking about. I don't believe it can be tested with that much accuracy. It's an easy solution, a very easy solution. A shortcut. Two different people could react to it differently. You can't say that you're testing the human mind and on the basis of that do this. People will undergo stress and parameters will go haywire . . . So do you understand how this machine works?' she asks me.

I have to admit, despite all my research, I really don't. Everyone has been so cagey about its inner workings. And from what I know about neuroscience, researchers are decades away from understanding how to pick memories out of a person's mind. In the UK, for instance, lie detectors are still not allowed in courtrooms because they are not considered accurate enough.

Following Sharma's conviction, Dere immediately lodged an appeal with the high court, complaining that the Brain Electrical Oscillations Signature test was bad science. Her eyebrows furrow as she thumbs through Aditi Sharma's case notes. 'She must have been under tremendous stress at that point of time at which she was made to undergo the test,' she says.

Six months later, Sharma was released on bail. That's where she is now, staying at her family home in the picturesque northern state of Jammu, near the Himalayas, far from the heat and noise of Mumbai.

'Have you seen her?' I ask.

'I just spoke to her father,' she tells me. 'He says she just wants to get on with her life.'

The slow Indian judicial process means that it could be many years before Sharma is back in court. 'Now that she's on bail it can take anything between five and ten years. Basically there are people in jail whose appeals from 2001 are being heard. You know, there are even old appeals from 1996 who are on bail that are being taken up now,' says Dere. This is common in Indian courts, where cases often drag on for decades.

Before I leave, she hands me a bunch of green and white papers, detailing the judgment from the case. This is what gave Mukundan and the forensic labs in Mumbai and Gandhinagar the licence they needed to run even more Brain Electrical Oscillations Signature tests on criminal suspects. The judge's remarks, stretching to ninety-three pages, read almost like the script of a romantic movie:

'This case presents a tragic scenario as the budding and flourishing love relationship between accused Aditi and deceased Udit Bharati, which was on the threshold of marriage and sailing smoothly with consent and approval of the parents on both sides got swerved and sank into tragedy . . . This is a case based purely and simply on circumstantial evidence alone.'

Later, when considering whether the Brain Electrical Oscillations Signature test should be counted as evidence, the judge accepts Sunny Joseph as a scientific expert: 'Thus the use of these tests both during the course of investigation and trial is held to be admissible.'

'Suddenly there's been a burst of these cases where the police have used lie detectors,' says Dere, shaking her head. 'I've always said that anything to do with the brain is more complicated than this. People have different stress levels. You can't follow something blindly,' she says before I leave.

That evening, though, I worry that perhaps I have been too dismissive of Mukundan's invention. If the brain is just a set of neurons and signals, isn't it possible that he *could* have invented a fairly reliable truth machine, even if other scientists don't know how it works?

Later I call up Geraint Rees, a professor of cognitive neurology at University College London and an expert on brain scans, for a final opinion. The human mind and human behaviour aren't simple things to read, he tells me. 'This is like saying let's find a machine that could find out the physical state of my computer,' says Rees. 'Let's say some imaginary device could read all the files on my computer just by measuring the activity on its case. Well I do not believe that we are even remotely close to having that kind of device technologically, and the brain is of course several orders of magnitude more complicated than the desktop computer.

'I have seen nothing in the history of brain imaging to say that we could ever get the degree of precision necessary to detect lies with complete certainty. My opinion is that the technology is still at the research level,' he adds.

Perhaps Mukundan is *too* geeky, I think to myself, as my time with the inventor draws to an end. He is so convinced that his truth machine works, and others are so keen to benefit from it, they seem to me to have ignored the potential for human error; that minds are more unpredictable and complex than computers.

The challenge for all engineers is balancing human uncertainties with scientific facts. Five years before the start of one of the world's biggest banking crises in 2008, for example, a former tutor in financial engineering at Columbia University in New York, Emanuel Derman, cautioned scientists hoping to make a career in the volatile world of finance that 'one must learn how to be neither too concrete nor too abstract, to choose some part of the spectrum between behavioural and quantitative, between science and psychology, between Feynman and Freud.' The key thing to realise was that they were dealing with people, not numbers.

But there was money to be made and the warnings went unheeded. When the recession started, quantitative analysts working on Wall Street – many of them maths and physics graduates – were accused of designing complicated software and calibrating it with financial models that were far too optimistic. Higher up, greedy bankers used the software without questioning it, and later, the system began to crumble. When banks collapsed, blame landed at the door of the unfortunate geeks. They had got it wrong, magazines reported. It was too late.

Here, the Gandhinagar forensics laboratory has wound down for the day. There are no more suspects due in, so the humming computers are switched off and the room is still. Mukundan and I are alone and I'm leaving soon.

'Man is not destined to be controlled by nature, but man is destined to control nature himself. This makes sense to you?' Mukundan says to me.

'I think so.'

'This is the big departure between a man and the creatures and the animals. They're part of nature, they're controlled by nature. Whereas human beings, we are destined to create a nature and then live in that nature. There's a huge difference between these two.' He pauses. 'But at the same time, I believe that somehow, I will not be able to duplicate a human brain. I don't know why I say that I'll not be able to duplicate . . .' he cuts himself off, mid-sentence. 'I may be able to make something close to brain, but I'm not able to conceptualise that possibility. I mean, if you do that, man becomes redundant, doesn't he? My reductionistic attitude makes me think that it is possible, but then it goes against my gut feeling. That strange gut feeling, it goes against it.'

Hearing this comes as a relief. I wonder if Mukundan acknowledges his place in that grey area between science and psychology, between Feynman and Freud.

'This is what interests me,' he continues, philosophically. 'That there is an organ here, there is a biological system which says that this is the purpose in my life. And you can define a purpose for your life. We all have different types of goal and we all struggle to achieve . . .' He tails off.

I ask him if he sees any conflict between his invention and the real-life complexities of the mind.

'They are not in conflict. They are not in conflict,' he replies. I'm about to interrupt, but he stops me.

'I found that there is no conflict!'

Driving out of Gandhinagar, I have never been happier to leave a place. My encounter with Mukundan has made me question, for one of the few times in my career, whether scientists and engineers are always a force for good.

But then in the end, Mukundan isn't really the problem. Indian leaders and police officers, in thrall to science and technology, seem willing to place their trust in new research and inventors, however wacky their ideas might sound to others. Maybe it's a side effect of the IT boom, which has created so much wealth and excitement about innovation. Indeed one of India's scientific strengths is that no idea is too off-the-wall to be entertained. Encouraging nuttiness can be a good thing, because it gives people the licence to think freely.

In the West, researchers sometimes complain that they can be too straitjacketed, with funding available for only certain kinds of work and too little scope to explore radically new theories. But even though these scientists may be less free, they do at least work in a rigorous system of peer review, which means that if research isn't published and verified, it usually ends up on the scrapheap. Good science tends to survive while the bad generally disappears. Parts of India, meanwhile, have yet to develop the full safety checks, balances and institutions to ensure that pseudoscience and crackpot ideas doesn't get out of control. It may take decades before the scientific establishment here is as organised and rigorous as it needs to be.

But this isn't just a question of time. Indian science is also a victim of the country's weak infrastructure.

I recall what the lawyer in Mumbai told me earlier, about the shortage of police officers, and trials being so slow that both criminals and the police officers get old and die before their cases come to court. Mukundan's truth machine probably would not be so popular if the legal system was faster and more efficient.

And the problems don't end with the justice system. Paying a simple water bill can take days. Many roads are falling apart, access to education, electricity and healthcare is patchy, and worst of all, corruption is rife. Thousands of India's scientists and engineers are working in conditions that are maddeningly difficult.

Outside the ivory towers of a few universities, the offices of the biggest IT companies and the best research laboratories, some parts of India can be as confusing and lawless as any banana republic. In fact a lot of the people I have met on my journey so far have complained that India's main obstacle to progress is that it's badly governed. It's the reason that Narayana Murthy, the boss of Infosys, once considered entering politics.

I'm still thinking about this on the way to the airport. I find it hard to believe that Indian brains, which are so imaginative when they're coming up with inventions to serve farmers and the poor, haven't also thought about solutions to this problem. Could the geeks have a way to bring more logic and order to how the nation is run?

CHAPTER 6

GEEKS RULE

The ground beneath our feet is always moving. Two hundred million years ago South America, Africa, India, Australia and Antarctica weren't where they are now but were part of one enormous supercontinent that spanned a whole hemisphere of the planet. The split happened around thirty million years later, after which they gradually drifted into the positions they occupy today.

When this violent breakup happened, gigantic volcanic ranges burst through the crust, while vast new oceans filled the gaps that the shifting continents left behind. The earth looked like a pie that had been torn into slices. At its edges, the land buckled and creased. In the north of India, the Himalayas were created. And when India's west coast finally broke free from the island of Madagascar, another one of these geological creases became the Western Ghats.

Today this 1,600-kilometre stretch of high hills blanketed in forests, is one of the most remote places in the world. There are at least 5,000 different types of flowers. Elephants, snakes, tigers, and cave bats all live here, secluded from the rest of India. Zoologists who have occasionally ventured into the Ghats have found new species by the handful. One time, they discovered a dozen multicoloured types of frog, including a species that was thought to have been extinct for a century. The mountainous territory is inhospitable. The second I cross into it, civilisation disappears. It's eerily silent in every direction, save a cry from the odd macaque. I can't get a mobile phone signal.

And yet here in the middle of the Western Ghats, in what can reasonably be called nowhere, I descend into a steep valley and find myself in one of the most advanced cities on earth.

I arrive in the evening, just in time to catch dinner at one of the first restaurants in this half-built metropolis. It's an American-style diner and I'm their only customer. There are pictures of Lucille Ball and Betty Boop on the walls and the floor is covered in black and white chequerboard tiles. Lots of dishes are off the menu today because it takes so long to fetch ingredients from the nearest city, Pune, which is two hours' drive away. I'm in the mood for corn, but they're out. So I ask for pancakes. Also out. I settle for an omelette, which comes served by staff shaking their hips to Elvis on the jukebox. They slide bottles of French's Yellow Mustard and Heinz Tomato Ketchup onto the countertop of my red leather booth. This could be New York in the 1950s.

'Pie, ma'am?' asks a skinny waiter in a uniform that matches the floor tiles.

'Alright, I could go for pie. Do you have the rum'n'raisin cheesecake?'

'We don't have any cheesecake, but the Key lime is nice, ma'am' he says, pointing to the picture on the back of the menu.

It's unnaturally gloopy and sickly sweet when it arrives, with a neon-green sauce squeezed over it. The odd taste reminds me of how far I really am from New York. Through the fuzzy light from the glowing signs over the window, I can see nothing but the dark hills.

This isn't *The Twilight Zone*. This is the city of Lavasa.

I first read about it in an advertisement in an inflight magazine, and became intrigued by what the advert claimed will be a metropolis governed mainly by machines. A bank of centralised computers, I read, will control everything here from household security to the transport network. It's a half-billion dollar project to build, from scratch, an urban dream in the middle of the mountains. Even the roads leading here have had to be carved out of the hills. It's the biggest thing to happen to the Western Ghats since the Cretaceous Period.

Standing on the promenade in the heart of Lavasa, I have a vantage point across the entire site. Ten years ago there was nothing here except a few tribal villagers living in low thatched huts. They grow food by terracing the slopes and waiting for the monsoon rains to feed their rice and vegetables. And now they can be found on the outskirts of Lavasa, watching this city rise from the valley like a girl gazing at her mother while she puts on her makeup.

If it looks surreal to me, it must look bizarre to the villagers. There are tall, thin, multicoloured apartment blocks in long terraces; they appear to have been lifted brick by brick from the Italian streets of Portofino. The opulent chalets above me, nestled inside the forests, could be from Bavaria. In the brochure, the Lavasa Corporation has used pictures of the quaint English city of Oxford to illustrate how picturesque Lavasa will look when it's finished. It's as if the developers have picked the most beautiful parts of Europe and transplanted them here.

Right now though, it's a ghost town. Apart from the occasional sound of construction work, the city is eerily quiet. One worker, on top of a long wooden ladder, is splashing water onto a wall to make sure the sticky concrete doesn't dry out too quickly. The American Diner, like Lavasa's half-finished homes and shops, borders an artificial lake that fills the dip in the valley. It's among the few places on the city's small promenade already up and running, to give the trickle of visitors – those brave enough to take the perilous two-hour journey through the Ghats, swerving around steep roads, with enormous boulders threatening to fall from the cliffs above onto their Jeeps – somewhere to eat. There's also a state-of-the-art hospital, which looks deserted. Electricity pylons stretch from here into the horizon, standing tall in the sun like marching aliens.

When I reach the canary-coloured town hall, I see men in suits and sleeveless yellow safety jackets standing outside for a smoke. It's the only building that could be remotely described as busy, but by any normal standards it's very quiet. The paint is fresh, some of the furniture still has its plastic wrapping on it, and every blade of grass in its gardens seems to be perfectly watered and of uniform length. This is the opposite of an Indian city.

'Indian cities have not distinguished themselves in the annals of urban management in terms of how well run they are,' agrees Scot Wrighton, the city manager for Lavasa, whose small office is upstairs inside the town hall. A wiry American with glasses and grey hair, he's responsible for running the city until it receives its first residents and elects a real mayor. Although this is an Indian project, the developers scoured the world for an expert who knew how to run towns with Western efficiency and cutting-edge technology. Wrighton was their choice.

He tells me he was trained in public administration at the University of Texas before helping to manage a few midwestern US cities. 'You've probably never heard of them,' he says. 'Wichita and Kirksville.'

I haven't heard of them, but I can imagine that travelling from the American Midwest to the Western Ghats must have been a culture shock for him.

'I'm just really excited to be here,' he continues. 'I love India and I work with some really bright, creative and intelligent people here at Lavasa. But our biggest challenge is that . . .'. He stops. 'This is being taped!' he coughs nervously, glancing at my voice recorder and letting slip a laugh under his breath. 'Our biggest challenge is that we're asking folks to think differently about government, and that doesn't come easily here in India because government operations are much more centralised here . . .'. He looks uncomfortable. 'When you first come to India, it's a little more chaotic than the West. Erm, that's not something I'm the first to say, but it's a . . . it means that what we develop here at Lavasa has to be sensitive to the Indian culture and the Indian systems, which are different.'

We both know what he's too polite to say. In the decades after independence, the Indian government wrapped so many miles of red tape around every state function that their style of bureaucracy became known disparagingly as the *babu raj*, for the empirelike stranglehold that small-minded public officials, the *babus*, had over people's lives. Twenty years ago, before the economy was liberalised, it could take months or sometimes even years to process a passport application or access land records. Although the bureaucracy has since relaxed a little, much of the government remains creakingly cumbersome to this day. And this infuriating system has filtered down to the streets.

Many Indian cities are unplanned and riddled with slums. Affluent districts have security guards on constant watch, or locked gates at least. Since 24-hour access to any kind of amenity, from water to electricity, is rarely guaranteed, people who can afford it have their own electricity generators and water pumps.

So the challenge for Lavasa's planners is to create a city that doesn't suffer from these problems. The way they hope to do it is by wherever possible replacing human bureaucrats with machines.

Miles from the reach of even the police and the emergency services, Lavasa is, by accident or design (I can't figure out which), forced to be self-sufficient. The chairman of the Lavasa Corporation, Ajit Gulabchand, dreams of turning this city into its own governmental entity, so it can do whatever any other Indian city is allowed to do, from providing healthcare and education to levying taxes. His ambitious promise is that Lavasa 'will be a city that governs itself' using technology, leapfrogging cities in the rest of the world.

But this isn't just an idealistic community. Lavasa is also a profitable real-estate development. Mumbai is only a few hours away. And the nearest city, Pune, is famous as an up-and-coming IT hub. In fact, the more I wander around the perfect pavements and delicate fountains in the blistering midday heat, the more I notice how hard they're trying to attract the kind of nerdy IT workers who are working in India's booming technology companies like Infosys and Tata Consultancy Services. There's a videogaming arcade opposite the American Diner. In the next few years, developers will also be building a space theme park, masterminded by the same people who created the American Space Camp in Alabama.

This may be India's first city designed for Generation Y. It's a geek's paradise. And not only will the geeks live here, the geeks will rule.

'Electronic governance is really nothing more than conducting the basic transactions of government via an electronic portal,' says Scot Wrighton. This means replacing paper-based filing, official forms and bills with digital ones, and transferring every point of contact between the government and its citizens online. The philosophy behind it is that automating the government can make bureaucracy faster, easier and more transparent. But the idea itself isn't new. It came about more than twenty years ago, when computers were becoming cheap and software more sophisticated. And then around a decade ago, technology caught up with the vision. Countries began putting these ideas into practice, using the terms e-government and Government 2.0.

The earliest nations to dip their toes in these silicon waters in Asia were Japan, Singapore and South Korea. They happened to be tech-savvy societies already. South Korea was so successful at turning itself into what it called 'Cyber Korea' that within five years it didn't need ten per cent of its government workers any more. In China in 2000, the Beijing government rolled out software called Digital Park, which allowed local authorities to process forms, statistics and finances online. And Taiwan was among the first places on the Asian continent to introduce a website that allowed utility bills to be paid online twenty-four hours a day. Nowadays every country on earth enjoys at least a little e-governance. The US, for example, spends 80 billion dollars a year on its IT systems.

Here in India, however, it being a nation infamous for its stiflingly complex bureaucracy, many assumed the idea of electronic governance would never stick.

Yet despite the poverty, the plethora of dialects and so many distinct, self-governed states, Indians took to the idea of e-governance almost immediately. Big technology firms like Infosys, Tata Consultancy Services and Wipro already had the technological expertise. They also had legions of highly trained, cheap staff. But most importantly, geeks like Narayana Murthy *wanted* to help. And so the desperate call to improve India's bureaucratic machinery was met with a nerdy solution.

In November 1999, in a move reminiscent of Jawaharlal Nehru's funding of India's space programme more than thirty years earlier, Andhra Pradesh – a large southeastern state that borders Maharashtra – launched the country's first e-governance project. It was a statewide network that electronically connected government offices to each other so they could communicate online. Next, the state installed hundreds of one-stop computer terminals where citizens could pay their utility bills, get birth and death certificates and licences, reserve bus tickets and download forms online.

The going wasn't always smooth. Some politicians in the state didn't like the system. Occasionally the kit crashed. Nevertheless, citizens were using it and so this first small experiment in Indian electronic governance worked. And then, after the turn of the millennium, came the IT boom.

Over time, Indian technology companies became better at designing electronic solutions to governance problems.

Here in Lavasa, one of the major companies responsible for installing and maintaining the technology is Wipro, one of India's big three IT firms. The linchpin of the e-governance system is a website through which residents will be able to pay their bills, access emergency services, report any problems, make complaints and do anything else involving the government's help. Households without computers will have a digital automation unit fitted in their homes to give them access to the site. The hardware will be replaced every four years or so and the software will be automatically updated through the Internet cloud. 'It's a slimmed-down, more efficient' infrastructure, Wrighton explains.

The Lavasa public relations team take me next to speak to the person from Wipro responsible for installing the hardware. He's known here only by his initials, UGK. He won't tell me what the U stands for but the GK means Gopal Krishna. 'Lavasa on a proactive basis would be looking at every aspect of infrastructure in the city,' he tells me, 'whether it is the streetlights, whether it is the roads, whether it is utilities. In the phase one, we would be having approximately 70 kilometres of optical fibre.' Metre by metre, researchers are mapping the city using a geographic information system. It includes water pipes, fibre optic cables, electrical wires, transport links, and the footprint of every building. If a pipe bursts, they will know exactly where it is.

UGK continues, 'You will have smart metering enabled which will allow you to capture the points of failure on a predictive basis, a preventative basis. It will also exactly pinpoint where the fault is, so that would help identify it quickly wherever the problem is. So all this would ensure that a resident at Lavasa would experience a very quick turnaround of faulty actions and repairs around that.'

I'm impressed, but at the same time I can't escape the feeling that I'm being given the hard sell. Indeed, from the slick brochures to the manicured gardens, it all feels like a giant sales pitch. But then, I guess I should have expected this. If the Lavasa Corporation doesn't attract a critical mass of at least a hundred thousand residents, there simply won't be enough teachers, doctors, lecturers, shop staff and other people to supply and use the services. It will remain a ghost town.

The PR team and the staff continue to drill me with the idea that Lavasa won't only work here, but that it can be a role model for the rest of India. 'We can't just cram more people into these already overloaded cities,' says Scot Wrighton. 'What we're going to have to think about is how to structure that and deliver those services differently. That's the laboratory of Lavasa. The vision of the chairman is that we can create a new governance model that can be replicated elsewhere. That's a terribly grand and idealistic goal, OK. It really doesn't exist anywhere else. So his idea is that we will be the laboratory, and figure out what works and what doesn't work.'

He suggests that I check out the corporate video in the building next door. He's in it, he tells me, half proud and half embarrassed.

It's as professional as a Hollywood movie. Over helicopter shots of the lush hills is a quote from Byron:

'There is pleasure in the pathless woods,
There is a rapture on the lonely shore,
There is a society where none intrudes,

By the deep sea, and music in its roar,
I love not the man less but nature more . . .'

Lavasa Corporation's chairman, Ajit Gulabchand, appears on screen in sunglasses and a sharp suit. '400 million people will migrate from rural areas to the urban areas in India over the next forty years,' he says, his thick, silvery hair fluttering as he walks past some bushes, the towering hills behind him. 'This huge migration took a thousand years to happen in Europe. It will happen in India in just the next forty years. India will have to expand its cities and towns.' The solution, Gulabchand announces, is Lavasa.

I go to bed that night, surrounded by the silent, dark hills, feeling as if I've arrived in Jurassic Park but the dinosaurs haven't escaped . . . yet.

The next morning, after a breakfast of dry toast (the valley is short on food supplies again) in Lavasa's chalet-style hotel, I ask for a tour of the entire 25,000-acre site. The woman from public relations agrees. Most of the site is empty land, with only the odd bulldozer lazily nuzzling the dirt. Reaching the edges, where the black Tarmac gives way to dirt roads, we stray into tribal territory.

From the viewpoint of the simple farms on the top of the hills here, the stark modernity of the European-style valley below is very clear. It looks impossibly futuristic and self-contained. The number of people living in Lavasa, the PR woman tells me, will be deliberately capped at 300,000 to make sure that services aren't overwhelmed. The city will be a quarter of the size of Mumbai but with only two per cent of the population of that city. There isn't a town on earth I know of that is so tightly controlled that the size is decided upfront, except maybe a retirement village.

I also wonder what this means for the poorer families on the outskirts of the city. When the Lavasa Corporation arrived in the Western Ghats, she continues, 150 families moved out of the valley.

'They just moved from their land? Didn't they mind?' I ask.

'They were hardly connected to the city,' she replies. 'There used to be a bus maybe once a day, maybe less.' Now there are regular buses. And to help keep the peace, the corporation also gave them electricity connections for the first time, and built crèches to educate the local children. 'This is better for them,' she insists.

'Are you sure?' I ask. In the distance I can see a long line of dirty tents where the construction workers are staying until the project is finished. On the roadside, a few people have built tiny roadside shacks with corrugated metal roofs, to sell the workers cheap food and drink. I wonder aloud what else they do for a living these days.

Sixty or so of the locals, she says, now work in a bamboo handicrafts factory that the Lavasa Corporation built to supply basic furniture to the houses in the new city. When I visit it, I find a warehouse swirling with bamboo shavings, and people working on the floor alongside empty Wipro computer boxes, which are being used to store the completed objects. There are other employment opportunities for the villagers too, I'm told. 'They can be useful in the city too. They can be your maids, your drivers.'

I leave Lavasa the following day. On the way to Pune airport, the driver asks me if I want to take the shortcut. My flight is leaving soon and we're already rushing pedal to the metal.

'Yes, take the shortcut. Hurry,' I tell him.

'Are you sure?'

'Of course I'm sure.'

He tells me that he's scared to go down the minor roads because people living outside Lavasa throw rocks at vehicles coming out of the city when they see them. 'I've got marks on the car already. Let's not take the shortcut,' he advises, missing the turning deliberately.

I wonder then whether electronic governance runs the risk of turning India into an even more split society, by introducing a digital divide where economic divides already exist. Lavasans (this is what I imagine the residents will be called one day) will be living their high-tech, sheltered lives parallel to the forest-dwelling tribes just a few mountains away.

'We'll make your flight, don't worry' says the driver, spotting the look of panic on my face.

I sit back, staring through the window. The question I ask myself on the last hour of the journey to the airport is whether the geeky governance model being used in Lavasa can be replicated and multiplied, as the corporation seems to believe. For it to work, it would have to meet everyone's needs, not just those of the wealthy and privileged. But the closer I get to Pune, the more unlikely it seems. This place, like many other Indian cities, is a sprawling mix of poverty and wealth. First I pass lopsided handcarts loaded with bananas, then misshapen office blocks with broken windows. Then come the potholed roads, scraggly beggars and queues of skeletal rickshaw-wallahs.

I think about Lavasa again when I'm back in New Delhi. The capital is chaotic, crowded and dirty (although not as bad as it was a decade ago, when tourists landing at the airport felt as though they were entering Dante's seventh circle of hell). Visitors either love or hate Delhi and, although I've personally learned to love it, in my experience more hate it. The city is crawling with grizzled men from distant villages trying to make a quick buck. Taxi drivers almost always overcharge and occasionally the haggling can get abusive.

In a few ways, though, it does have parallels with Lavasa. The 'new' part of New Delhi is a planned city and is constantly renewing its network of housing and shopping quarters. There used to be huge slums tucked away on empty plots of land here, forcing everyone to live cheek by jowl, but many of them have been demolished by the local government in the last decade and replaced by purpose-built homes at the far edges of the city, next to the Yamuna river and too far to commute into Delhi. So the slum-dwellers who used to make a living by begging, cleaning or selling magazines on the streets have little to do these days but play cards and drink themselves to sleep.

Pushing the poor to the margins hasn't solved Delhi's problems. I've been warned against venturing out alone at night. The crime rate is high enough to have encouraged families to build more and more gated communities, usually patrolled by layers of security guards. The city is a patchwork of so-called 'colonies', like military cantonments, with names including Friends Colony and Defence Colony. Rich Delhiites often seem oblivious to the poverty on their doorstep, an attitude that can appear confusingly callous to foreign visitors. But this is the story of modern-day India. As salaries rise, the middle classes retreat from the craziness of the cities, insulate themselves in air-conditioned cars and huge homes, and try to forget that the poor exist. India is full of these kinds of gated communities. In Mumbai and Bengaluru, they take the form of giant, portered apartment blocks, while in roomier cities like Delhi they look like this – vast, patrolled colonies with beautiful gardens, sports clubs and mini mansions.

So despite Ajit Gulabchand's idealistic dream of replicating Lavasa all over the country, it's clear to me that building isolated e-governed communities can't solve the nation's underlying problems of overpopulation and inequality. It is, as the corporate video says, a 'society where none intrudes', like a gated community taken to its obvious conclusion. The poor majority still needs somewhere to live, whether it's a shack on the edges of Lavasa, underneath a Delhi flyover or a sprawling slum in the heart of Mumbai.

The realities are even messier than they appear on the surface. A recent study by Transparency International, for example, revealed that all households below the poverty line in India together pay more than 200 million dollars in bribes every year, just to get access to basic services like education and utilities. Nearly half still bribe police officers. The poor are the least likely to have access to utilities like electricity and water.

If it's possible to draw a thread of order out of this chaos using electronic governance, then that solution probably has to come from the central government in New Delhi. So my first stop is the Ministry of Communications and Information Technology.

It's in a pink and cream stone building that has been appropriately named Electronics Niketan, meaning House of Electronics. I never enjoy going to government departments. Indian civil servants are famous for turning up to the office only a little before midday, taking long lunch breaks, and knocking off at about four. So as a journalist, they've become synonymous in my mind with long queues and frustrating interviews. Occasionally they're even dangerous – I was once bitten on the foot by a poisonous spider while waiting to see an official (ironically) in the Ministry of Health.

This time, though, my assumptions turn out to be wrong on all counts. The ministry's e-governance unit is an efficient nerve centre. And there are no dangerous spiders in sight. The civil servants working here are busy assessing computer-based management projects around the country and deciding which ones work and which should be scrapped. There are freshly printed glossy posters on the wall advertising the National e-

Governance Plan. One shows a woman wearing a *bindi* shaped like a curly 'e'. Another has a photo of a man from Rajasthan with a long moustache, both ends of which have also been curled into 'e's.

From the corridor I can hear a woman shouting. 'Do you want to work here for three years or not?' she demands. There's a long pause until, 'Yes,' a girl's voice says, quietly. I catch a glimpse of her walking meekly out of an office back to her small orange cubicle.

The bark belongs to Vineeta Dixit, the principal consultant on the e-Governance Plan. She's a no-nonsense civil servant who has worked in the private sector for more than fifteen years, which has proved long enough for her to learn what the realities are like on the ground. She invites me into her office almost immediately, rather than keeping me waiting as I was expecting.

'Historically the government has not performed a hundred per cent on all fronts. I don't know what the reasons are,' Dixit tells me, her brow furrowing. 'The government did fail in its basic duties in ensuring every child or every citizen had access to basic education, potable drinking water, you know. All those issues, they are very real issues. But it's also true that governments in such countries, like ours, have access to limited resources.'

And this is why, she continues, the government has turned to IT companies and private sector workers like her for a solution. Electronics Niketan may be the nerve centre for e-governance, but the actual software design, hardware installation, operation and maintenance are done by the same technology giants that I met back in Bengaluru, including Infosys and Tata Consultancy Services, as well as other big players including Wipro and the mobile phone company Reliance. Dixit herself is a geek, who studied social policy and development at university, with a special focus on the role of information and communications technology. Experience has taught her that the private sector can deliver services in countries like India more cheaply, efficiently and far faster than can the slow, old-fashioned bureaucrats.

To this end, the department has a vision statement. It's one of those peculiar management-speak mnemonics that are supposed to remind everyone what their jobs are about. And it says Indian e-governance should be SMART:

> Simple
> Moral
> Accountable
> Responsive
> Transparent

Some of the projects may turn out to be a waste of time and money, Dixit admits, but on the whole they're delivering what Asian governments dreamed they might deliver twenty years ago, at the start of the e-governance revolution. 'For us, e-governance equals disintermediation,' she says. In other words, cutting out the middle man. It offers a way for ordinary people to interact directly with the government, bypassing the *babus* and their *babu raj*.

'There have been lots of projects happening all across the country. A lot of good stuff is happening on the ground. There are several examples,' Dixit says, flicking through a stack of papers. 'I mean we could start talking about what happened in Andhra Pradesh more than ten years ago. We can talk about what happened in Karnataka. Then you have that small project in Kerala. And you have e-*gramin* [*gram* means village] in Gujarat . . .'.

It sounds as though all the Indian state governments are silently digitising themselves.

Dixit continues, 'One of the issues has been that each of these places has limited resources, in terms of money available to them, in terms of manpower available to them, in terms of access to technology or access to private sector who would implement this project. So the government of India has been contemplating for a while to figure out how to stitch all of this together.'

And that's where the National e-Governance Plan comes in, to monitor the projects and help provide funding. Dixit's job isn't easy. 'We're creating a digital service delivery infrastructure. We have statewide area networks, which provide a minimum of two megabits per second of Internet connectivity up to the block level,' she says. There are around 100,000 blocks across India, each one of which is a collection of half a dozen or so villages. The plan is to have at least one computer or kiosk connected to the statewide online network, known as a common service centre, in every block.

'What's the cost of all this?' I ask, recalling how expensive it has been to introduce similar hardware in Lavasa.

'We're looking at an estimated cost of about six to eight billion dollars. For everything,' she replies. It's going to become mandatory for every state government to spend two to three per cent of its annual income on IT.

This is more than I expected. It means that e-governance isn't just a marginal effort to tidy up the Indian bureaucracy; instead it's a wholescale attempt to change the way India is run.

While eight billion dollars is a huge sum of money in India, this is still small compared to the amount spent on many Western public sector IT projects. When politicians in the UK, for example, wanted to computerise the National Health Service's paper medical records, they estimated it would cost them more than 18 billion dollars, and that was for a country of just 60 million people, not more than a billion citizens like in India. In the end the idea was shelved because it was working out too difficult and expensive. Dixit laughs when I remind her of this example. 'I don't know why the NHS ran into problems,' she says. 'We face the same set of challenges. For example, we've been digitising land records for a good twenty years now, OK.

And these records, some of them are over two hundred years old. You can't even touch that paper on which the record is made because it would fall to pieces.'

Dixit's team is achieving so much with relatively little cash by exploiting India's strongest resource: 'It's the people, it's all about the people,' she says. Indian computer engineers don't just design and build software more cheaply, the experience they have in maintaining applications for governments around the world means that they also have an unparalleled bank of knowledge about what works and what doesn't. This has helped to make them among the most efficient city rulers in the world. It's the same reason why the super-geeky city of Lavasa is being built in India and not somewhere else.

And it's also the reason why, within a decade, this geek nation could be more computerised than some of the most advanced economies in the developed world.

The Indian Income Tax Department, Dixit continues, has had a centralised website since 2002 and every passport office became computerised in 2007. According to the Ministry, running these offices online has made the system both faster and fairer. Now, proportionately only half as many people as before pay bribes when they're filing their taxes. Wherever there has been a common service centre, waiting times have fallen by between 20 and 40 per cent, and each person has saved on average 60 to 110 rupees, depending on the state. This is only one or two dollars each, but it represents a huge saving for India's poorest families.

The priority now, adds Dixit, is to reach out to the remotest corners of the country. The ministry is devoting more time to e-governance projects in the picturesque southern island territory of Lakshadweep; in the small states on the northeastern border with China; and in Jammu and Kashmir to the north, where India is fighting a war with Pakistan. The scheme has been so successful that most states, including these ones, are on track to be electronically governed within a few years.

So far, there are 70,000 common service centres up and running in India. In three months, there will be 100,000. And in three years, there will be 250,000. At that point, even the tiny local leadership bodies, known as *gram panchayats* (village councils), will be hooked up to this giant electronic government mainframe.

At the weekend I take a road trip through Rajasthan, which is half a day's drive away from New Delhi. I've been told that this state has one of the biggest IT departments of any local government in India.

On the dusty golden tourist triangle between Delhi and the Taj Mahal, Rajasthan seems an unlikelier place for an electronic revolution than the lush hills of Lavasa. It's a state with roughly the same population as the UK, except much of the terrain is desert. My destination is Jaipur, the capital city, and the route takes me through sand-coated plains, interrupted only by the odd hill and parade of camels. Arriving in Jaipur, I see rows of small shops selling semi-precious gems, silver trinkets and heavily embroidered bridal gowns. Once upon a time the shop buildings were painted red, but they have now faded to a peaceful pink, giving the impression that the whole place is suspended in a warm cloud of candyfloss. Rajasthani men wear red turbans that dwarf their tiny faces, and the women wear flowing red and yellow two-piece *lenghas*, with their veils draped over their heads. I feel as if I've arrived at a wedding.

Underneath the pretty façade, however, this place has its own surprisingly geeky history. When the Maharaja Sawai Jai Singh founded it around three hundred years ago, it was India's first planned city. There are nine sectors, to match the nine planets in the ancient astrological zodiac, with the shops also grouped in multiples of nine. And it's home to one of India's five ancient astronomical observatories, which includes a 27-metre-high yellow sundial.

The nearest e-*mitra* kiosk (*mitra* means friend), I've been told, is in the government hostel around the corner from an electronics store on the long MI road. But on the way there, despite all the positive statistics, I can't shake a lingering scepticism I have about e-governance, especially in smaller cities like Jaipur. For one thing, most ordinary Indians are not online. Fewer than one in a hundred, for example, have access to a broadband Internet connection. Secondly, India's bureaucracy seems far too cumbersome to be tamed by computers. Can the old *babu raj* really be abolished by the geeks or will it divide India down the middle, creating a two-tier society of technological haves and have-nots, like inside and outside Lavasa? The government hostel is an intricate yellow stone building with small white windows and giant green gates. Inside it hasn't held up as well as it has on the outside. The walls are dark and yellowing, faded tiles litter the floor, bars block the windows and tatty brown chairs are lined up against the wall. Unlike in most banks and ticket offices in India, however, there are no writhing, sweaty queues. At one end is a row of counters, like those in a bank, each with a flatscreen Wipro computer. There are young men, each with long sideburns and wavy, oiled hair, sitting at two of the screens.

'It's been like this for five years,' one of them tells me. He's Sanjay Sharma, a 22-year-old trained to operate these computers and process people's bills (his job required fifteen days of training). 'Before this there were tickets, licences, registration. Now there's software on the computers,' he continues. He reels off a list of which bills customers can come here to pay. It includes water, electricity, housing loans, IDEA mobile phone and Rainbow telephone bills.

A man comes forward to pay his electricity bill. Sharma types a PIN number onto the screen, calling up all his details. He takes a cheque and gives him back a printed receipt.

'It seems pretty fast,' I comment.

'It's taking thirty seconds only,' he replies, smiling. Before the computers arrived, this single bill might have taken hours to pay, and the teller probably would have taken a bribe.

The whole of Jaipur is sleepy on a Sunday, and made more dreamlike still by the soft pink palaces and the lumbering, painted elephants ferrying up to the surrounding forts. My husband and some friends are in the city for the annual book festival, which is being held under colourful canopies nearby. So I have a choice. I can join them and the foreign literati while they sip tea and discuss India's history, or I can head to the Rajasthani Department of IT on the slim chance it might be open.

I choose the geeks.

I'm relieved to find the department open. There are two young computer engineers busy scribbling in one office. They're part of a team of a hundred programmers here who are helping to build the state's e-governance software. In another room, next to a large sign telling me that 'Information is Power', is R K Sharma. A bespectacled computer analyst and programmer, he's the person in charge of all the e-*mitra* kiosks.

'We have maxi-mum IT professionals in government sector. When I started we had sixteen persons working here. One. Six. Now we're having four hundred persons,' Sharma tells me proudly. He spends so much time designing computer programs that he speaks in peculiarly clipped sentences, each one as functional as a line of code. They're half words and half numbers. I can see pigeons walking on the dirty windowsill outside Sharma's office, while inside, it's stuffed with papers organised into bundles tied with red thread. They are the unmistakeable mark of an Indian government office. One by one, though, these bundles are moving online.

'This computer is a fantastic tool. I'm monitoring the entire project sitting at my desk!' he says. Sharma is busy but he so rarely gets visitors that he opens his laptop enthusiastically to show me the state's e-governance website. 'Today is Sunday. I can tell you what is happening in the state today.' He clicks. 'Let's see. All districts.' Click. 'Five hundred persons are operating five hundred kiosks across the state. Every kiosk has a broadband connection and they are doing their job simply sitting at their home. In Alwar district since morning, this person he has collected one mobile bill. Telephone bill, he has collected twenty. Electricity bill, he has collected twenty-two. In Alwar district, thirty-two consumers have been served, with this much revenue.' He points at the screen. 'This is only Alwar district. Today, Sunday is a holiday, but 6,141 consumers have been served since morning. In collection, twenty-nine lakh rupees.' A lakh is one hundred thousand.

I had arrived in Rajasthan with a sneaking suspicion that e-governance might be an expensive but doomed government shortcut. A white elephant. But Sharma is convincing me otherwise.

He offers me another example. 'If I sell a property to you there are a lot of legal formalities, we call them legal documents. Deed, we call them deed. This deed is very big. And the deed has the entire details of the property, the terms and conditions I'm selling to you. So government sub registrars they were doing this manually,' he explains. In 2000 his team started a pilot project in Jaipur to digitise these hefty land records. 'We made entire system IT enabled. Then the outcome was the time. Earlier property registration was taking seven days, to deliver these documents to citizens. After this, it comes for only one day. Same day. Same day, we are delivering this document to public.'

But convenience isn't the only reason that the system has been so popular. 'Every government office, some corruption is there,' he explains. 'In some states there may be higher, in some states there may be lower, but it is there. So what we are doing is providing the entire detail of fees on website. Entire details, how much he has to pay government. So computer increased transparency. Right information to public,' says Sharma.

And this means people are paying fewer bribes than before. At the e-*mitra* kiosk in the city, anyone who comes to pay a bill can't be cheated because the amount they owe is printed automatically onto their receipt. The Ministry of Communications and Information Technology has found that, in five out of the ten states that had computerised their land records by 2008, bribes have either reduced significantly or been completely eliminated.

So far in Rajasthan, the e-governance kiosks extend only as far as the larger towns and cities. But when the programme is completely rolled out, there will be more than 6,000 kiosks serving every village in the state. One of the other beautiful things about the project, Sharma says, is that most of the operators will be women. 'In Jaipur division, twenty-two women are already working,' he explains. 'This is helping women empowerment.'

On Sunday night I start the long drive back to New Delhi, stopping halfway at a dimly lit hotel to get some buttered toast and watch the evening headlines on a tinny TV set.

One story has been dominating the news for the last few weeks. It's a crime that took place two decades ago, but which is only now being heard in court. A 14-year-old girl called Ruchika Girhotra claimed that the director-general of police for the state of Haryana – Shambhu Pratap Singh Rathore – had molested her when he was giving her tennis lessons. What should have been a simple police matter turned into a saga of coverups and intimidation. Witnesses said that Rathore harassed both her and her family. Her brother was mysteriously picked up and tortured by police officers for a crime he didn't commit. And then Ruchika's school expelled her. According to the newspapers, politicians spent the whole time shielding Rathore from prosecution. Tragically, the girl committed suicide a few years later.

Photographs of Ruchika, wearing a purple sweater and with her hair tied up, have been appearing in the newspapers next to snaps of her suspected molester, Shambhu Rathore. He looks like a villain in a cartoon, his long moustache twisted into a constant grin for the paparazzi. 'Wipe off his smirk,' reads one headline. Cameras circle him on his way into court as he walks through the angry crowds. One onlooker, the news reports say, has taken the law into his own hands and slashed Rathore's face with a knife.

When the verdict arrives, it turns out that 67-year-old Rathore will spend six months in jail. The public are outraged at the leniency of the sentence. But what surprises me more, my eyes still glued to the television screen, is that it has taken nineteen years for the courts to hear Ruchika's case in the first place.

Unfortunately, this case isn't the only one of its kind. Lawyers and forensic scientists in Mumbai had warned me that the Indian legal system was slow. A combination of police corruption, a shortage of judges (there are 3,000 unfilled job vacancies in the courts) and a backlog of cases means that justice here is not so much a promise as a probability, depending on whether the suspect survives long enough to reach the dock. A tally in the summer of 2009 found that there were 52,000 cases still waiting to be heard in the Supreme Court in New Delhi. In the smaller high courts, there were four million, and in the trial courts around 27 million.

While electronic kiosks may be speeding up the financial machinery of government, justice remains one of India's biggest problems. Speedy and fair trials lie at the heart of a functioning democracy. Indeed, the government's SMART vision statement – Simple, Moral, Accountable, Responsive and Transparent – means nothing if it can take two decades for the courts to hear a child molestation case. But the challenge is immense. The Indian legal system remains the final lingering throwback of the *babu raj*. In fact it's buried under even higher mountains of yellowing paperwork than it was fifty years ago.

On the cold, dark drive back to New Delhi I wonder if it's even possible to invent electronic grease powerful enough to oil the wheels of the creaking Indian judiciary.

There are thirty-six courtrooms in the Delhi High Court, and every one of them is drowning in paper. There are heavy case reports, legal books and court judgments everywhere, piled shoulder-high in the arms of advocates and spilling out of every cupboard and shelf. We simply can't get through cases fast enough, explains one of the judges, Justice Ravinder Bhat. The high court gets 65,000 cases a year, he says. 'Our strength, the number of judges we have, is currently about forty. That's the biggest challenge we face. A murder case could be disposed of in three or four years, but an appeal can take double that.'

The difference between this court and others, though, is that it's unusually progressive. Delhi High Court was the one to famously legalise gay sex in India. Recently it also gave women in the Indian Air Force equal status with their male colleagues.

And now, in an unprecedented move, the court officials here have decided to get rid of all the paper.

'I have seen those days when we used to type with typewriters, and if there is some correction in there, we had to retype again! Files can be 2,000 pages long, very big,' says Girish Kumar Sharma, the registrar who's in charge of this huge operation. A hairy-eared man, he has worked here since 1978, when there were only ten courtrooms. The paint is peeling off the walls in his office and thick, dirty blinds sit heavily on the windows. Over time, the court has accumulated hundreds of thousands of files. But instead of keeping them in the enormous basement underneath the high court building, Sharma has begun converting them into electronic documents, clearing up space as the court expands. His team has digitised more than 35 million pages already.

'We have scanning and we will start e-filing also, and when we start e-filing, then the scanning will be less,' he says. By e-filing, he means that the paperwork associated with active trials will go straight online. Nine e-kiosks have been fitted throughout the high court to give lawyers access to a central website that contains all the documents. Until everything is paperless, radio-frequency identification tags have also been attached to some paper cases, so people can track where they are in the building.

But the biggest changes are happening in the courtrooms themselves. Outside, next to a sign asking people not to chew and spit betel-leaf juice onto the floor, the foyer looks like a stock exchange. Every court official has been given a biometric ID card to beef up security at the entrance. Hundreds of dark-suited men and women are staring at a tall LED screen, where cases and courtrooms are matched up to one another in giant red letters.

The screen flashes. Case A26 is in room 24. There's a flurry and a man in a black gown disappears behind a door. I follow him. This courtroom is different from the others. It's the location of an experiment into electronic justice. Sharma calls it an 'e-court'. Here, high court officials are trying to figure out whether computerising cases can help speed up trials themselves.

Indian courtrooms, being colonial leftovers, look a lot like British ones. The proceedings are in English and the protocols the same. The main difference between the two is the thronging crowd. I squeeze into one of the stained, scruffy white chairs to the left of the wood-panelled room, next to a queue of lawyers, each of whom is clutching thick brown files and waiting for his or her case to be heard. Among the legal books lining the room, I spot one oddly titled *Spectroscopic Properties of Inorganic and Organic Compounds*.

'Seven!' reads out the clerk, taking a sip of water. There are two clocks on the wall, one analogue, the other digital. The judge's bench is empty save a huge computer screen, which holds copies of the proceedings. Instead of lugging legal papers around, he also has everything on a pen drive. He reads out a date from the monitor, which the lawyer scribbles down before leaving.

'Eight!' says the clerk.

The judge taps his screen and drags his finger along it to the next case. It's a domestic dispute dating back around five years, when a house was apparently split between different members of a family. The grandmother has brought a complaint.

'She's not even allowed in the living room,' says a lawyer in a black turban.

'These locks were in place even in the passing of the status quo order . . . It starts at page 35,' says the opposing lawyer.

'Which paragraph?' asks the judge, scanning through the file on his computer.

'Erm . . . Para 17, page 44,' he replies, shuffling through a fat beige file, before turning to a stack of books tied with blue thread that lie on the table next to him. The lawyers' half of the courtroom hasn't had screens installed yet, which means that advocates are relegated to using paper files. 'The living room is in my possession,' he says at last, the judge already ahead of him.

'So you're saying half the property is in your possession?'

'But the mother can't even entertain guests,' pleads the first lawyer.

After some deliberation, the judge makes his decision. 'She can have the drawing room. Let there be a situation for the common areas, let both area's parties have the keys . . .' he says. 'This is a 90-year-old. If this lady is in discomfort, it is our duty . . .'

The case is finished within a few minutes. A chubby girl in red lipstick transcribes the judgment.

'Nine!' says the clerk.

In half an hour, Justice Bhat hears six cases. The judgments are uploaded onto the Delhi High Court website within a day. Most of them are property disputes like this one, over unauthorised construction or land rights. A26 is a case in which someone built an extra storey on his property, apparently without permission. By half past midday, all the cases have been heard and the bustling room is almost empty.

The e-court system is much quicker, Justice Bhat tells me, taking off his thin-rimmed glasses to give his eyes a rest. 'You get to handle your caseload very fast. I definitely feel there is a greater speed for us, though lawyers may take their own time.'

In a year, the registrar Girish Kumar Sharma tells me, he hopes to have entirely computerised two e-courts so that both judges and lawyers can use the new system. And when this happens, he believes the case backlog will shrink even more quickly. Indian justice, in this court at least, will be speedy.

Living most of my life in Britain, I've always had a slight suspicion of IT. The promises that were made in the 1990s, that offices would disappear and we would all have computerised workspaces at home, that queues would vaporise as everything went online, never really materialised. Digitising our lives is harder than it sounds. In fact failed IT projects have reportedly cost the British government more than 39 billion dollars. A computerised scheme to hand out farm subsidies cost 530 million dollars alone and is already 'obsolete', according to newspaper reports. Plans to issue UK citizens with electronic identity cards were also shelved when it turned out the project would be too expensive.

And so from my vantage point in London, the idea of digitising even a sliver of India's vast, maddening bureaucracy seems almost ridiculous. And yet somehow it's happening here faster than in many other more developed nations. Perhaps it's because technology is cheaper, perhaps it's down to the fact that there are so many computer engineers, or maybe that the civil servants are just particularly geeky, but the Indian government is succeeding where others have failed. It's already on target to deliver a biometric ID card to every adult within a couple of years. The scheme, incidentally, is headed by the former CEO of Infosys, Nandan Nilekani.

Walking out of the courts, I recall something that Narayana Murthy told me at the Infosys offices in Bengaluru. 'Any system that heralds the death of distance and that makes access to governance anytime, anywhere, is likely to make life better for the poor people, whether it is in the area of healthcare, whether it is in the area of land record systems, whether it's in the area of getting drivers licenses for cars or vehicles or whatever it is, all these things benefit tremendously from technology,' he had said. 'So we need it, we need it.'

'But it's quite a logistical feat,' I had replied, my British scepticism kicking in.

'It is, yes, absolutely. It is. I'm an optimist and I think it will happen. It may not happen within my lifetime but it will happen in yours.'

CHAPTER 7

THE IMPOSSIBLE DRUG

I spend the next day in bed with a fever.

'Just take some honey in hot water and a paracetamol,' the pharmacist says. It's a harmless virus that's spreading through New Delhi, he adds, brought on by the biting winter cold. It lasts only a few days and I'm near the end of it. But I decide to take some extra advice from my last appointment in the city, Professor Virander Chauhan at the International Centre for Genetic Engineering and Biotechnology. He's one of India's foremost experts on infectious disease. A friend of mine, hearing about my scientific quest through India, had told me he might be able to help in my research into that other famously geeky Indian pursuit – medicine.

'We're full! We're *full* of infections!' he growls as I sniff into a tissue.

'You name any infection, except for sleeping sickness and maybe yellow fever, and we have it in India,' he continues. 'My suspicion is that there are also lots of unknown viruses lurking in India, Bangladesh, Malaysia. So if you are in Delhi, and every time you get a fever or something, every time the doctor just tells you that you have a viral infection, nobody knows what the virus is. Mostly in four or five days your fever comes down, but we still don't know what you had.'

Chauhan, now white-haired, did his doctorate at Oxford University in England, before returning to India. And I get the sense that working here for so many decades, after living in a place that's relatively infection-free, has given him a grumpy cynicism about eradicating infectious diseases. 'I know of one girl,' he says. 'She was from my driver's family, and she had an infection and got gangrene. By the time they got her to a hospital in Bihar, she was dead. You will hear these stories if you live in poor countries.'

He pauses, glancing at my nervous, still-feverish face with his good eye (he has glaucoma in the other). 'You'll be fine.'

In India, he tells me, the greatest risks to most people's health are not so much the mysterious infections, as the ones that doctors *can* identify. They include diseases like malaria, polio, dengue fever and tuberculosis, which have almost disappeared in the West. In India, for example, two people die of tuberculosis every three minutes. It affects two billion around the world, mainly in developing nations.

Tuberculosis is a particularly smart bacterium, says Chauhan, causing fever, fatigue, weight loss and a characteristically persistent cough. Its genius lies in the way it spreads. It's carried through the air in the tiny, phlegmy droplets that patients breathe out. If they're not treated, an average sufferer infects between ten and fifteen others a year. To make things worse there hasn't been a new medicine to treat tuberculosis for forty years. When the disease began to vanish in Europe and the US, Western pharmaceutical companies gradually lost interest in inventing new drugs to treat it. So now it takes a cocktail of old drugs to fight it; none of them could beat the bacteria alone, but even as a cocktail they don't always work.

'Why are there so many new drugs for treating cholesterol? It's because they are Western diseases,' he explains. Infectious diseases like malaria and tuberculosis have been largely ignored.

'The West is quite besotted with how to make people live

longer or look better. All their focus is on hypertension, diabetes, how to lose weight and so on,' he says. Decades spent working in India's rural areas have taught him that fighting infectious diseases, 'viruses, bacteria and suchlike', is not a battle that doctors here are likely to win. It's a war of attrition: mankind against billions of bugs.

'So there are no Indian scientists looking for a solution? New drugs?' I ask, thinking back to how innovative scientists had been in Lucknow and wondering why Indian researchers wouldn't be trying to beat diseases like tuberculosis too.

'There is something,' he admits. 'But I don't think it will work.'

This takes me next to the government tuberculosis clinic in Chennai, a quiet port city hugged by the Bay of Bengal in the southern state of Tamil Nadu.

I've always been scared of hospitals. It's the smell, the sounds, and the feeling that I could catch something I didn't come with. And Chennai happens to be full of hospitals. In a nation famous for its doctors – after all, more than a fifth of the doctors working in Britain's National Health Service are of Indian origin – this city is known as India's medical capital. Western health tourists flock here every summer to have their hearts bypassed and stomachs sucked at rock-bottom prices. A new knee costs as little as $5,000, including a private room for ten days, in a skyscraper clinic between shopping malls and slick apartment blocks.

The government tuberculosis clinic is different. For one thing, it's in Chetpet, which was once a village and is now a sleepy district in the heart of old Chennai, dating back to when the city was still called Madras. Centuries ago this area was a busy trading post for the East India Company. The British built churches, railway stations and courthouses in a flamboyant mix of Gothic and Hindu styles; many of them are still in use today, converted into museums and administrative offices. The people, though not quite as old-fashioned as the architecture, are still wearing 1960s-style nerdy spectacles with heavy black frames and bottle lenses.

I find the clinic at the end of a messy driveway flanked by overgrown trees, opposite a stagnant lake walled off with barbed wire, and just down the road from a roadside shack selling questionably cheap X-rays. There's something ghostly about the building. Standing outside, it reminds me of a colonial hospital from the same era as Florence Nightingale, when places like this weren't so much expected to make people better as provide a last home for the dying. The white marble walls and floors have been scrubbed so hard that cleaning fluid has seeped into every pore, and now a vapour of it permeates the building.

Tuberculosis is believed to be the oldest infectious disease in the world. Each civilisation has given it a new name: the ancient Greeks knew it as *phthisis*, while a hundred years ago in New York and London it was called consumption. It killed Chekhov, Kafka, Keats and Napoleon. Even Egyptian mummies dating back 3,000 years have been found to carry evidence of the disease. So maybe this otherworldly sanatorium, shrouded in a bleach mist and ringing with the faint, ghoulish echoes of rasping coughs is exactly where a disease as ancient as tuberculosis belongs.

Paradoxically, though, it also happens to be the site of some of the most cutting-edge scientific research into the disease.

The place looks deserted. There's no furniture in the foyer, and not a scrap of colour save a poster on one wall warning patients to cover their mouths if they cough. I flinch, catching the sound of someone emptying his lungs. The scary thing about a tuberculosis cough is that it doesn't come politely from the throat; it comes in unashamedly loud, phlegmy waves from every corner of the torso, sometimes for minutes on end with no let-up, spewing showers of bacteria into the air.

The sound of the cough is giving me an attack of hypochondria. I waver here for a few minutes, unsure whether to continue or leave in the nearest taxi. I dive into my bag, furiously looking for a wet wipe to clean my hands (although of course this is little defence against an airborne disease). And then I hear footsteps.

An elderly woman in a red sari hobbles down a wide marble staircase, a giant window at the top silhouetting her tiny frame. 'What are you looking for? Are you a patient?' she asks, covering her mouth with the drape of her sari.

'No, I'm looking for the Tuberculosis Research Centre,' I tell her.

'It's behind the clinic.' She nods at the door and shuffles away. 'This clinic is for patients only.'

The cloud of bleach extends to the Tuberculosis Research Centre. But unlike the clinic, it's a new building, busy with white-coated doctors and scientific researchers. I'm meeting Dr Sujatha Narayanan, the deputy director of the Department of Immunology. She's a world expert and I should be concentrating on my questions now. Instead my imagination is running away with me. I'm picturing myself trapped in the cold white clinic, being tended to by Tamilian matrons in crisp white bonnets like Florence Nightingale.

Tuberculosis research can be risky, admits Narayanan, a small, middle-aged woman in a red and white cotton sari, wearing a small nose ring and green bangles. She's doing little to calm my nerves.

Narayanan has worked here for twenty years, devoting her life to understanding tuberculosis and spending much of that time in this dark office, surrounded by cracking blue paint and rusty filing cabinets. 'I mean it's a very, very difficult organism to deal with. People are scared to grow the organism in ordinary labs.

That's why in our centre there is a specialised lab where culture can be done,' she says.

There are two physical features that make tuberculosis such a difficult disease to cure. First, it multiplies so slowly. Unlike other types of bacteria, which divide into two every fifteen to twenty minutes, making it easy to get big samples quickly, a tuberculosis bacterium takes a frustratingly sluggish twenty-four hours. This means that it can take weeks to grow a culture in a petri dish that's sizeable enough to study and run experiments on. The other problem is the structure of the bacterium. Some bugs are easy to combat because they're easily overpowered by a general antibacterial medicine, but tuberculosis has a fat, waxy outer wall that makes it extraordinarily tough. Often drugs can't reach it; they just bounce off.

'Is it true that one third of the world has tuberculosis?' I ask Narayanan.

'Yeah, one third,' she says. 'See in India, what happens is that by the age of twelve, when children are given a Mantoux test – that's a skin test, where you inject a tuberculosis antigen into the skin to see how the system is responding, the immune response – well, by the age of twelve in India, most of the children are positive. That means they are already exposed to tuberculosis. But that doesn't mean we develop the disease. It's sitting there.' Millions of Indians have tuberculosis bacteria resting latently in their bodies, not reproducing. The disease only becomes active and starts causing symptoms when they become malnourished or so ill with another disease that their immune systems can't fight off its effects, she explains. So about a tenth of all the people who have tuberculosis bacteria in their bodies will actually become sick.

Scientists have spent futile decades studying the disease, trying unsuccessfully to find a way to disable it. As Professor Chauhan told me, it is so widespread that eradicating it is now almost impossible. The best that researchers can do is understand what makes the bacteria so virulent. Downstairs at the research centre, for example, they keep a repository of around 3,000 different samples of sputum (a mix of spit, mucus and phlegm) from tuberculosis sufferers across south India, each of which has been harvested over the last seven years and then cultured in a laboratory to help doctors learn more about the disease.

'Every day two hundred, three hundred sputum samples are arriving at bacteriology,' says Narayanan.

Now all I can think about are these sticky, disease-ridden tuberculosis globules, maybe only metres away from my mouth. Statistics run through my head as I make a frantic estimate of how likely it is that I might have already caught it. I watch Narayanan. She looks sickly. Maybe it's her age, my hypochondriac mind speculates, or maybe not.

'Have *you* ever had tuberculosis?' I interrupt as she's talking.

She smiles and looks down at a polished conch shell on her desk. 'No, but actually you know in my family, my great-granduncle had it, and my mother-in-law had it and my uncles had it. I thought I had it once.' There's a pause. 'It was fifteen years back. I was doing some research and needed some cells to culture them and infect them with tuberculosis to see what are the effects, the immune response and everything, you understand?'

'So you used cells from your own body?'

'Yes. We had no volunteers so I subjected myself to the test. It's one where you put a tube into the lung. We have lots of tuberculosis patients, but no normal people to test, so myself and my students subjected ourselves to that. And when we did it, when we cultured my sample, we found there, one bacteria growing. Only one.'

Narayanan suspected that the tube that had been down her throat was contaminated with a tuberculosis bacterium. There was a chance that she had accidentally infected herself.

And so she took a course of medicine. The treatment they use in the clinic, on the recommendation of the World Health

Organization, is called DOTS, which stands for Directly Observed Treatment, Short-course. A medical supervisor is supposed to monitor the patient for around six months (it used to take a year and a half, but the treatment regimen has improved), dishing out the appropriate drugs until they've recovered.

'Because of that, you know, I gulped a lot of drugs. I gulped them for eight months,' she says. She escaped tuberculosis, but the long and draining treatment gave her another illness. 'I came down with diabetes,' she tells me. 'I already had the diabetes gene, no doubt, but the latent diabetes was triggered by those drugs.' There's only the slightest tinge of regret in her voice.

I remember the story I was told at school about Marie Curie, the legendary chemist who spent her life studying radioactive metals, only for them to end up fatally poisoning her. Endangering your life for science is the kind of hazard that only those truly passionate about science would face. Narayanan's far braver than I am, and a true geek.

She continues. The DOTS programme has a 95 per cent success rate when completed, she says. This makes it pretty effective. But because the treatment is so slow, many patients stop turning up to the clinic part of the way through, assuming they've been cured when in fact they're still infected. 'Once the cough reduces, they stop the drug,' she says. The bacteria survive, infecting more people, and becoming stronger.

And this is giving rise to another problem: Narayanan is now seeing some patients on whom DOTS doesn't work at all, because their particular strain of tuberculosis has evolved to beat the effects of the drugs. These new strains, which first emerged in the 1980s, are known as multi-drug-resistant tuberculosis and extremely drug-resistant tuberculosis. At the last count, in 2008, almost half a million people around the world had the multi-drug-resistant kind of bug. Half of these were in India and China.

But the biggest victims nowadays are sufferers of HIV, the deadly virus that weakens the immune system and therefore makes it harder to fight off the tuberculosis bacterium as easily as healthy people can. Sometimes, Narayanan tells me, HIV patients who have been cured of one strain of tuberculosis will catch another strain in the clinic's waiting room. Of the 1.8 million people who died of the disease in 2008, half a million had HIV.

'You know, because of HIV and because of the extremely drug-resistant tuberculosis, new drug development is very urgent in the whole world. We have to have better drugs,' says Narayanan, shaking her head. 'New diagnosis and vaccine, that's the hope, but it's a long, long way off.'

Making medicines is an expensive and difficult business. It can cost up to a billion dollars to create a new drug from start to finish, including the failures. And there have been a lot of failures. Occasionally pharmaceutical companies spend decades researching something that seems as if it might work, only for it to fall through at the last hurdle, in clinical trials.

Many have lost interest in finding a tuberculosis treatment altogether. This is partly because patients in countries like India tend to be less able to afford medicines. Since drug creation is expensive, every pill needs to fetch a high price to justify the investment and the sums don't always add up. Inventing cures for infectious diseases like tuberculosis and malaria requires years of expensive research and is unlikely to bring back high financial returns. As Chauhan told me in Delhi, the focus has instead switched to lifestyle illnesses that strike the rich, such as heart disease and obesity.

But the failure to find new medicines isn't just down to money – tuberculosis also happens to be a particularly tough adversary for medical researchers. When the last drugs were found for it, in the 1950s and 60s, it was through little more than luck. Scientists started with existing molecules, which are the basic bunches of chemical elements that make up natural things, or with compounds, which are larger bunches of molecules. They judged how promising each one might be depending on its shape and characteristics, and then tested out the likeliest ones on animals and then on humans.

Isoniazid, for example, is one of the compounds now used in the DOTS programme, was created in 1912 but it took doctors another forty years to learn that it could treat tuberculosis (for a while, it was used an antidepressant). They derived another tuberculosis drug, Rifampicin, from a new species of bacteria found by chance in a pine arboretum in France in 1957.

After the 1970s, this hit-and-miss approach to drug discovery was gradually replaced by high-throughput screening. It's technically more sophisticated, but the principles are the same. Chemists pump millions of potential molecules into a plastic grid, which are rapidly tested and analysed by robots, automatically spitting out the promising ones at the end. It remains to this day the most popular way of developing new medicines. And it has certainly made the early steps of drug discovery quicker. But in some ways, the results are still a lottery.

And it means that scientists need to come up with a new way of doing research. Since all of the obvious molecules and compounds that might work against tuberculosis have been tried, Narayanan is among a growing cohort of medical researchers who believe that an entirely different approach is needed. Professor Chauhan, in Delhi, doesn't think it will work, but one increasingly popular method is genomics, the science of how living things work at a fundamental level – in their genes. And according to some scientists, it's the best hope for a new tuberculosis drug.

I'm sharing a small, sweaty room in New Delhi with ten thousand fish. Each one is as small as an anchovy, iridescent, with dark, horizontal stripes.

'They're zebrafish,' says 38-year-old Dr Sridhar Sivasubbu, watching me through a filled tank, the fish flitting between us like a shimmering curtain. He's one of a team of researchers at the government-funded Institute of Genomics and Integrative Biology who are trying to unlock the secrets of animal genes. This emerging branch of biology promises to teach them more about the human body and how it reacts to diseases like tuberculosis.

Like a lot of Indian laboratories I've visited, the institute is where you'd least expect to find it. When I turned up to this part of New Delhi's industrial zone, Naraina, I assumed that I'd come to the wrong end of town. The journey took me through a dirt road, squeezing past painted trucks, bursting with their cargos of freshly printed newspapers and long steel rods destined for construction sites. Children from neighbouring slums were playing in the mud in the streets. I passed a man coughing mucus into the gutter – it seemed that the haunting sound of tuberculosis had followed me here. Yet this institute is at the cutting-edge of Indian medical research.

'Zebrafish is actually a very good model to study a lot of human disorders,' continues Sivasubbu. He's crazy about fish (he has a PhD in fish genetics). 'From an evolutionary point of view, fish are the first animals with backbones, and progressive evolution has brought us all the way to humans. So if you ever want to study a simplistic set of genes that are present in simplistic settings, then it's gotta be zebrafish. They have most of the organs, blood, brain, nervous system that we have.

'We could work on other animals but a fly doesn't have blood, for example. Mouse is very good but doing lots of experiments on mouse is always going to be expensive. So we thought we'd train our hands on zebrafish.'

The training is almost done. Just recently his team of ten researchers at this unlikely institute made headlines by becoming the first in the world to sequence the zebrafish genome.

A zebrafish's genome is a list of all the information inside its DNA – the thin, long molecule that influences how it's made (just like the fruit genomes that the biotechnologists were tinkering with in Lucknow to create their long-life banana). It contains the instructions for the proteins that are the building blocks of all living cells. Each step on a ladder of DNA is made up of a pair of chemicals, and these chemicals are represented by two letters of the alphabet out of A, T, G or C. Sequencing a genome means figuring out which chemicals are on which step of the ladder.

When this is done, scientists are left with a kind of biological computer code, looking something like this: ATG TCC CTG GCC GGC. The human genome has around three billion of these letters in its DNA. The wild zebrafish, even though it's only a simple minnow, has 1.7 billion.

These letters by themselves, though, are meaningless. The real reason why sequencing a genome is important is that it helps identify the specific genes in a DNA string, each of which is thousands of letters long. Every gene corresponds to a particular characteristic and behaviour in the animal. The human genome contains around 23,000 genes. In comparison, a zebrafish genome is more basic. But since we evolved from vertebrate fish-like creatures, for research purposes the zebrafish genome is almost like a very, very early human ancestor. So sequencing a zebrafish genome and comparing it to a human one can tell us exactly which fundamental genes fish and humans have in common. As a way of figuring out how our bodies work at the most primary level, the zebrafish makes a great model.

Besides its simplicity, another benefit of experimenting on zebrafish is that each one is almost identical to the next. This means it's easier to make general observations and construct a genome for the creature. Every small, blue-capped tank in the institute's facility contains around a hundred fish. 'Each one hundred of these fishes are essentially clones of each other, because they're all siblings,' says Sivasubbu.

It's getting uncomfortably hot, so we leave the zebrafish facility for Sivasubbu's office. The institute is like a maze, or maybe like a prison, with low ceilings and dark walls. There's a guard at the door, clutching a rifle (they hired him because shortly after moving into this laboratory thieves walked in and took some electrical cable). We pass a girl in a Benetton sweater sitting in a corner underneath a stack of hard drives, whose job is to check that the computer programs are working correctly. Then we go past a locked room containing a giant genomics analyser, a machine that makes the institute's work possible.

In 2003 the Human Genome Project, an international effort based in the US, finished sequencing the human genome at a cost of almost three billion dollars. It took them more than a decade. But this groundbreaking work pushed manufacturers to make sequencing equipment that's so fast and cheap that now commercial companies can process millions of pairs of genetic letters in a day. They can sequence a person's genome and burn it onto a DVD for as little as $20,000. It was thanks in part to these genomics analysers that in 2009, Sivasubbu's team became the first scientists to spell out all the letters in the genome of a wild zebrafish. They followed up soon afterwards by sequencing the first full genome of an Indian person (an unnamed, healthy 52-year-old man). It was a landmark achievement.

'The first time we operated it, we did it without training,' brags Sivasubbu with a wide smile. The team likes to play up its successes because their funding depends on them, but I get the impression they also feel that they have something to prove. 'We couldn't afford to wait two months for training. We didn't have any support, because the tech guys were over in the US. Here you have to become a master of your equipment yourself, be able to fix it at three am in the morning. You have to come up with innovative ways of fixing it. Now, look outside. Why are we in this crappy environment? It's because it's India. We don't worry about our surroundings, we just do it.'

I love their make-do-and-mend nerdiness. But despite the boasts, the truth is that sequencing a genome is not nearly as difficult nowadays as it used to be. Instead, the real hurdle lies in understanding what it means. This is known as annotation: breaking the genome's code to figure out exactly how all those As, Ts, Gs and Cs translate into genes and how these genes correspond to biological characteristics.

Annotation is a far slower job than sequencing. One way to do it is by comparing genomes to find out how each differs from the next, and from those of other species. This often yields surprising results. In 2005, for example, US researchers studied a mutant strain of zebrafish that has a golden colour instead of the usual dark, silvery hue. They discovered that this golden colour corresponded to a specific gene, which also accounts for about a third of the difference in human skin colour between dark-skinned Africans and light-skinned Europeans.

Using this kind of approach, the potential insights are limitless. It might even be possible to find the human genes that make some people more or less susceptible to catching tuberculosis, or particularly receptive to certain drugs.

But making medicines is more complicated than just identifying letters in a string of DNA. Knowing that a gene exists doesn't necessarily help to find a molecule that can cure a disease. So even though genomics is becoming a bigger part of drug discovery, there's lingering doubt about whether this method can really work. At the International Centre for Genetic Engineering and Biotechnology, Professor Chauhan had warned me that genomics research had been hyped up since the start of the Human Genome Project, about thirteen years ago, but has in fact delivered only a few drugs. And these have been mainly for cancer, not for common infectious diseases like malaria, polio and tuberculosis.

On the plus side, studying a person's genome can often help to predict whether they might be more likely to develop diseases that are influenced by their genes. For example, a history of heart disease in an individual's family can make them more susceptible to developing it too. In theory, spotting the genes responsible for this trait early on should allow a patient to alter his or her diet or exercise pattern to lower the odds of getting sick later in life. Infectious diseases, however, rarely have a hereditary element.

The truth about the researchers who study genomes, Chauhan told me, is that 'they overstate their case, otherwise nobody would fund them. Really. Ask yourself what new drugs have been discovered for malaria, tuberculosis, and the answer is right in front of you. The answer will be none. The answer is not genetics.'

I have his words in my mind now, as I wander around the institute. I haven't seen a single person here older than 40. They're Generation Y. And each one is an evangelist for this new approach to medicine. But I can't help thinking that Chauhan was right. They're nowhere near inventing a new tuberculosis drug, or any other medicine for that matter.

The problem is that scientists still don't have enough data about all the genes in human DNA. Sequencing one Indian man's genome is no use when it comes to drug discovery, because there's nobody to compare it to. Scientists may need to sequence thousands, maybe even millions, of genomes to be able to spot the tiny genetic differences between people that could help identify why some catch diseases like tuberculosis more easily than others.

'You're right,' agrees Sivasubbu when I put this to him. 'In genomics, it's very strange. You cannot generalise things based on one genome. Ever!'

The research is still in its early days, with teams around the world amassing terabytes of data on more humans, different fish and other animals. Elsewhere in India, I've heard of one team of scientists that's sequencing a buffalo genome. 'There are several genomes that are ongoing,' Sivasubbu murmurs. 'I'm not sure about it yet, but I've heard rumours that one of the birds is being sequenced. Not sure what bird it is. Then there was a rumour about a goat.'

'This is a huge jigsaw puzzle that has to be put together,' he admits. There's a hint of hopelessness in his eyes. It's the look of a person who knows what he wants to achieve, but may not live long enough to see it happen. In the minutes just passed, dozens of Indians have already died of tuberculosis. The race to cure the surviving feels endless.

The problem with genomics is that however fast and cheap the machines are becoming, the odds of producing, any time soon, a new drug effective against a tough, infectious disease like tuberculosis remain wafer slim. Even if it could work, researchers don't have enough information yet. And what they do have is scattered like puzzle pieces in laboratories around India and across the world.

One piece of the puzzle is back in Chennai, with Sujatha Narayanan at the Tuberculosis Research Centre. Unlike the team at the Institute of Genomics and Integrative Biology, who study humans and fish, her job is to understand the other end of the disease – the tuberculosis bug. 'All strains of the bacterium might behave the same when they're cultured in the lab, but underneath this, genetically there are differences,' she explains.

These differences have turned out to be more astounding than she anticipated. In 2006 she published a study with researchers in the US, South Africa, Germany, the Netherlands, France and Switzerland, looking at different strains of tuberculosis in each of these countries. Their research showed that the bacterium has been evolving in such a sophisticated way that it tailors itself to the local environment. Every region appears to have its own particular strains of the disease. The Indian subcontinent has seventeen types, South Africa has five, Europe has thirty-five and the US – that gigantic melting pot of immigrants – has collected 207 strains.

'We have found that the bacteria which is located in one region of the country has different genetic characteristics versus bacteria which is there in Europe, or the US, or Philippines, or in Malaysia or Singapore. The bacteria and the host are co-evolving,' says Narayanan. Some strains, the team discovered, affect only certain ethnic groups. When it first arrives in a new country, for example, 'the Indian strain does not infect an American, the Chinese strain does not infect a European,' Narayanan explains.

'In a San Francisco study – you know, San Francisco is a very cosmopolitan city where people from various countries come and live there – we took sputum cultures and from there we took DNA and analysed the bacterium that is present. It was a detailed study that followed people for eleven years. And it showed that the Indian people who are there, even though they interact with other people, the bacteria from them do not spread immediately to people from other countries. In the first and second generation of the disease, the Indian tuberculosis strain tries to infect only the Indians.' This means that strangers to the local strain could be, temporarily at least, safe from the disease.

Narayanan thinks that genetic material from the tuberculosis bacterium is mixing with genetic material from the people it infects, creating bugs that are better suited to that particular population. 'Indian bacteria in that region, they're used to a certain environment and the food they eat. They're evolving with the host,' she explains.

They don't yet know exactly how this happens. Pinning down what triggers genetic behaviours inside both bacteria and humans is 'a million-dollar question. We don't know these bacteria well enough to know exactly *what* they are getting used to,' she adds.

But it could be one clue and one step towards developing a new and more effective vaccine or medicine against tuberculosis. For example, scientists recently developed a new HIV drug using similar kinds of insights. They found that a tiny proportion of people in the world have a mutated version of a certain gene called CCR5, which makes them naturally resistant to HIV. This taught them that CCR5 must be involved in the infection – it began, they found, when the virus locked on to this gene. An American pharmaceutical company developed this research into an effective new medicine, known as maraviroc, which inhibits CCR5 and effectively blocks the route the virus would normally take into the human body. This new HIV drug was approved for use in the US in 2007.

In the Madras Institute of Technology, inside the leafy campus of Chennai's prestigious Anna University, a biologist has another piece of the puzzle.

Dr Sulagna Banerjee is a young-looking and bespectacled 37-year-old in a cream and red salwar kameez. The first step in her research has been trying to figure out exactly how tuberculosis infects people.

'The tuberculosis bacterium has a very sugary, waxy kind of

body,' she explains, calling up a picture of it on her laptop. I can see a wiggly red string, representing the sugary covering on the outside of the bug, stretching towards and then attaching itself to a green spring, representing the sugars on the surface of a human cell. 'The connection between an infectious organism and a host cell is mediated by these sugars. This is the first stage of an infection,' she explains.

Banerjee is trying to identify the sugars on the surface of the tuberculosis bacterium in the hope that, like the discoveries around the CCR5 gene in HIV infection, she might be able to figure out the pathway of the bug into the body. 'If you really understand the sugars and the linkages, if you identify them, I think that's going to be quite significant in the whole research field,' she says.

Unfortunately the laboratory she works in isn't safe enough to allow her to work on actual tuberculosis bugs. Tight laws governing research into dangerous bacteria and viruses mean that any work of this kind can only happen under controlled conditions. In this room, the paint is peeling, the wooden doors are flung wide open and the benches are overflowing with blue-lidded containers and cardboard boxes. Two dirty white lab coats are hanging on the end of the bench, where a couple of students are poking at something in a plastic tray. All the time, mosquitoes are buzzing in and out. One nips me on the leg.

So instead Banerjee studies the bacterium from afar, using computer models and less dangerous bacteria similar to tuberculosis.

Her bug of choice is *E. coli*, which is the bacterium that very often causes food poisoning. *E. coli* is easier to work on than tuberculosis. Not only is it safer, it also multiplies much faster, doubling in quantity every fifteen minutes. There's also just about enough overlap between an *E. coli* genome and a tuberculosis genome to make comparison possible. Like the zebrafish, it's a model organism.

Banerjee demonstrates for me how her experiments work,

taking out a small blue plastic box of the kind that Indian jewellers use to store costume earrings (in fact I suspect that she might have got it from a jeweller in the first place). Inside is a clear gel. Different parts of the cells in a bacterium have a positive electrical charge, which means they get attracted to the negative pole of a battery. And this is the characteristic that Banerjee uses to separate the sugars in the bacteria from one another. The sample is liquidised and this extract pulled through the gel using electricity. Since the smaller bits travel faster than the bigger ones, everything gets separated depending on how heavy it is. Next, the gel is stained with a blue dye, creating barcode-like blue bands, so she can see the different cell fragments. Each line is weighed using a molecular-level scale known as a mass spectrometer, and then gauged against a library of small molecules.

Banerjee has found three sugars so far. The next job is to identify other molecules that might block these sugars before they attach onto the surface of human cells and start infecting them. If it works, this could be the basis of a new drug.

But it's tougher than it sounds. There's a high chance that what works in a laboratory test tube might not work in a real person. Even if she does find a promising molecule, there's a risk that hitting the sugars on the surface of the bacterium might also accidentally knock out some sugars that are essential for the rest of the human body. What Banerjee is looking for is known in medicine as a 'magic bullet' – a treatment that hits only the bug and doesn't cause any side effects.

Her biggest problem, though, is that she doesn't have all the equipment she needs – or the safety clearance – to do this work alone. When she needs data that's specific to tuberculosis, rather than *E. coli*, she's forced to ask for help from other laboratories. 'See when you're starting out in research like I am, you have to compromise on what you really want to do,' she says apologetically, glancing around her lab as though she wishes it was as tidy and well-equipped as the American labs in Boston where she did her postdoctoral research. 'I would really love to do this sugar structure thing all by myself here but I can't because I don't have the resource for doing it. But if you kind of accept that it's not about *my* doing it, but getting it done, for the sake of it, then it's fine. I try to do the samples and stuff as far as I can do, as far as my lab permits. Not every university here is as well equipped.'

The same problem exists across India, where talent is plentiful but resources are often harder to come by. Between them, though, Narayanan, Banerjee and Sivasubbu are uncovering the separate clues that collectively could help create the new tuberculosis drug that has evaded Western scientists for fifty years.

'Indian scientists are resourceful. They're kind of open to the idea of getting it done,' says Banerjee. 'If you want to do it yourself, it never gets done. That's where we strike a compromise. We say OK, I'm willing to dilute my credit a little bit if it comes to publication, but at least let me get my work done. That's how we think of it.'

The thing is, their separate pieces of the puzzle need to fit together to work.

Owning a scientific idea can mean the difference between fame and obscurity, fortune and penury. And that's especially true in the medicine business. Every tiny molecule is patented and defended by armies of lawyers. The biggest drug companies are like scientific fortresses.

And that's why, most of the time, scientists don't want to share their puzzle pieces. It's the only way to protect their ideas. And it has been this way since the earliest days of modern science, when to ensure that the right person got recognition for a discovery, research was kept tightly under wraps until it was published or patented. In fact some of history's most vicious feuds have started over scientific credit. The nineteenth-century inventors Nikola Tesla and Thomas Edison, for example, quarrelled bitterly over their different methods for producing electricity. What famously became known as the War of Currents began with Edison launching a dirty campaign to convince people that his idea (which he'd invested a lot of his own money in) was better, but ended with him losing to Tesla when it turned out to be worse. Even now, scientists often find themselves racing against one another to publish first, knowing that if they're even a day later than their rivals, their work could be left close to worthless.

'Can we break out of these four walls?' asks Zakir Thomas, a government scientist at the Council for Scientific and Industrial Research in New Delhi and the person hoping to bring the pieces of the tuberculosis puzzle together.

The council has an illustrious history, dating back to the early days of Indian independence when Prime Minister Nehru was its president. It's still one of the most productive scientific bodies in India, funded by the government (the visual evidence of this is the queue of white Ambassador taxis, resembling fat London taxicabs, parked outside – the Indian official's preferred ride).

Government workers are usually sticklers for tradition, but Thomas's corner of the building is different. There's a spirit of enterprise about his office that reminds me of the innovative software companies in Bengaluru.

'Every turn of the wheel is a revolution,' says the screensaver on his desktop. He's a smiley, cleanshaven guy, wearing a grey zip-up sweater. And like Narayanan, Banerjee and Sivasubbu, he's dedicated to finding a cure for tuberculosis. He seems to have prepared an entire presentation for my benefit. 'Science has kind of bypassed this disease and unless we bring in serious science to look at it, we will not combat it. Where are the tuberculosis cases?' he says, pointing to a map. 'These are the documented cases. It is concentrated in southeast Asia. It is there in Africa. And so what is the market of this disease? The global market is around 300 million dollars. Now, the pharmaceutical industry's own estimate of drug discovery varies from 800 million to a billion dollars. So if you are a pharmaceutical company or a shareholder of a pharmaceutical company, would you let your board invest in this disease, where your whole market is so poor it's only worth 300 million dollars?

'Drug discovery is a highly risk-oriented, highly research-intensive business,' he continues. 'There's a huge amount of risk and a huge amount of fieldwork, and to recoup that, you give patent protection, and that industry and the person who undertakes that research get a monopoly of about twenty years to recoup on that investment. But when there's no market, the traditional concept of drug discovery fails. When this whole market, from which you're meant to recoup this investment, doesn't exist, the whole chain breaks.'

He flashes up some more statistics on the screen for me. 'If you compare the number of drugs in the pipeline for tuberculosis with drugs for cancer and cardiovascular diseases, the number of drugs for tuberculosis is only six,' he says. 'Cancer we can see there are three hundred and ninety-nine, and a hundred and seventy-eight companies are working on it. Cardiovascular, there are a hundred and forty-six drugs, by eighty-two companies. If you're putting ten drugs in the pipeline, one may come out if you're really lucky. The chances, with six drugs in the pipeline, of a real drug coming out for tuberculosis, is minimal. So the reason why this disease remains neglected is the pharma industry is not investing money into this. It's very simple. The crux of the problem is there.'

And that's why Thomas thinks the four walls around drug discovery need to come down. Scientists working on the big infectious diseases affecting the developing world, including malaria, typhoid and polio, have reached crisis point. Western pharmaceutical companies aren't investing enough money or time in developing a new tuberculosis drug, he says, and there isn't a single Indian company with the resources to do it alone.

So here in the Council of Scientific and Industrial Research, they're trialling a radical new idea. They believe that by sharing all the existing and upcoming research on tuberculosis, they might be able to come up with a new drug faster and cheaper. Their novel approach is to ask scientists all over the world to make their research available to each other for free.

On Thomas's noticeboard is a multicoloured poster:

OSDD Hackathon 2010
Are you a Geek?
Help us with the *Wave* to find
new drugs to fight Tuberculosis.
Open Call for Students & Software Developers
to create Next-generation Apps for
Collaboration & Annotation on the *Wave Platform*

Thomas is the project director of OSDD, which stands for Open Source Drug Discovery, the council's huge collaborative research effort into tuberculosis. The 'Wave' mentioned in the poster is Google Wave, a free, experimental software platform designed by the giant search engine company, which lets large groups work together and share data over the Internet. And the Hackathon is a coding competition for engineering students to develop a set of online tools that might allow biological scientists to use Google Wave to collaborate for research.

Thomas's team wants to bring all the pieces of the tuberculosis puzzle together so that researchers in different parts of Chennai will be connected to researchers here in Delhi, and to others elsewhere in the world. Open Source Drug Discovery promises to break down the barriers around research and create enormous, open pools of data. 'We're trying to bring together a platform which will create Science 2.0,' he says, beaming.

If this sounds like language borrowed from the world of computer science, that's because it is. Open source has its roots in the nerdy world of software engineering. It all started when the first commercial computers came on the market in the US in the 1960s. At that time, hardware and software were the same, so if you bought a computer, it came ready programmed with its own software. That all changed in the 1970s when the first separate software packages appeared, allowing users to choose which programs they wanted to use on which machine. Like the pharmaceutical industry, though, their investment required a return. Unlike hardware, software is easy to share. So to recoup their costs, companies kept their code private and locked the programs so that each user would have to buy his or her own copy.

But in that decade, a countermovement emerged. Open-software campaigners argued that programs should be freely available so that they could be used, adapted and improved by anyone. Renegade computer scientists developed Unix, one of the first free computer-operating systems. It proved such a huge success that developers continue to use it to this day, constantly tinkering with it to make more Unix-style operating systems, including the widely used Linux.

The success of open-source software cemented its place in the technology landscape as a tool for creating things just as well as, if not better than, private companies could. Computer scientists remain the world's best collaborators.

Nowadays some of the most active open-source developers in the world are in India. Being inherently free, it's one way of making products that are affordable for poorer consumers. In 1999, for example, a group of Indian engineers invented an open-source computer, called the Simputer. It was a cheap, handheld alternative to personal computers and has already been used in a few Indian e-governance projects.

The Department of Information Technology, where bureaucrats are implementing their bold e-governance policy, has its own cell dedicated to developing open-source software. They've created an operating system that supports eighteen Indian languages and is being used by the Indian Navy. It's the 'numero uno choice for supporting government and educational domains,' according to the government's website. Since then, the Indian government has also funded research into the world's cheapest open-source computer. In the summer of 2010 the tablet-shaped device, similar to the Apple iPad, was completed, reportedly available at a cost of only $35.

'Now we're trying to apply the same philosophy of open-source software to drug development,' says Thomas. But while it may be open, it's not exactly. They have about twelve million dollars of funding from the Indian government so far, which will rise to thirty-eight million in the future. It sounds like a lot of money, he admits, but it's just a fraction of what a pharmaceutical company would need to come up with a new drug. And it makes OSDD one of the largest projects in the world to attempt open-source medical research. There is a similar group based in the US, called the Tropical Disease Initiative but, he tells me, OSDD has more participants.

The team started off using a wiki, similar to the online encyclopaedia Wikipedia, which lets anyone change and add entries by registering and logging on, as the main portal for researchers and students to plug in their data. The next step is to pick out the molecules that could lead to potential drugs. Other scientists within the OSDD team will screen and test them.

But it all hinges on getting enough expertise together. So far, Thomas tells me excitedly, 2,700 people have registered on the website, from fifty-three different countries (when I call him a few months later, it's gone up to 3,800 in ninety-three countries).

Included in this mix are Sujatha Narayanan and Sulagna Banerjee, back in Chennai, and Sridhar Sivasubbu in Delhi, as well as university researchers and government-funded institutes in Lucknow, Pune and Chandigarh.

The project has created such buzz that private companies have also begun offering their support. The IT giant Infosys in Bengaluru, for example, is developing the software that will let OSDD researchers link their data over the Internet. Using a Web 3.0 idea known as the semantic web, the meaning of numbers and datasets will be encoded into the text, meaning that related bits of information will automatically link to each other.

'Infosys is developing that portal totally free of cost, because we made a presentation before Narayana Murthy,' says Thomas. 'We showed him the statistics of people with tuberculosis, and he just said, "We're in".'

In some ways, owning ideas is a peculiarly Western concept. A good example is the tussle over the evergreen tree neem. Native to India, its long leaves and bark have treated insect bites and cured bad skin for centuries. With proven medicinal properties, its twigs are still chewed to clean teeth instead of toothbrushes, and its oil is added to soaps and shampoo. For these reasons, in India the tree is considered sacred. But around a decade ago, an American agricultural company went to the European Patent Office to claim rights over a product that exploited the antifungal properties of neem oil.

Thousands of Indians protested. Neem, they argued, had been used in India for so many centuries that it was unfair for any company to claim it for itself. If the patent were granted, they feared it would leave other traditional Indian remedies open to ownership claims by multinational companies. Unlike modern medicines, they believed, these traditional remedies belong to everyone and should remain free.

After a protracted battle, the patent was revoked.

India remains remarkably patent-free even to this day. Recently, the Indian Ministry of Science and Technology tried to introduce a new piece of legislation called the Protection and Utilisation of Public Funded Intellectual Property Bill – similar to one that has existed in the US for decades – which threatens to put patent protection around all public research so that discoveries made in government-funded laboratories can be sold and commercialised. But again, Indian scientists have campaigned against the bill, calling it a barrier to free and open research.

So it seems appropriate that an idea like OSDD would have emerged here in India rather than anywhere else.

According, however, to Dr Hiroaki Kitano, the director of the Systems Biology Institute in Japan, OSDD has more in common with Eastern ways of thinking than just being patent-free. Kitano is a systems biologist. Technically, a biological system is a group of living things that works as team. For example, the circulatory system is made up of the parts of the human body that pump blood, such as the heart, blood, arteries and veins. Our nervous system carries information using the spinal cord, brain and nerves. Systems biology is a young branch of science that takes this concept and applies it to research. An illness, systems biologists believe, can't always be treated like a broken cog in a machine; rather, it's part of a holistic system.

The idea emerged in the mid-1990s, around the same time as computing power became plentiful and cheap. The reason for this was that crunching enormous amounts of biological data takes a lot of processing strength. One of Dr Kitano's day jobs, for example, is working at the computer labs at Sony, the electronics giant, where he helped build the AIBO or artificially intelligent robot. He is now also a key member of the OSDD team.

He tells me via email from Japan that there's something about systems biology that resonates with the ideas behind traditional Eastern medicine and that perhaps this is what has made it particularly popular among Asian researchers. Chinese herbal remedies and Indian ayurvedic medicine, he explains, are 'highly context-dependent.' While they're rarely backed up by the rigour of modern science, the useful lesson they do offer is the value of treating illnesses by looking at the person as a whole, not just the specific part of the body that happens to be sick, or only the bug itself.

Similarly, OSDD doesn't make use of a single branch of biology in isolation; it takes a wider look at the problem by drawing on different research from all the sciences. In practice this could mean, for example, taking Sulagna Banerjee's research on the sugars on the surface of tuberculosis bacteria and putting it in the context of Sujatha Narayanan's work into why some people get tuberculosis more easily than others, and then developing a new drug by also looking at Sridhar Sivasubbu's genomics research to figure out which medicines work best on which populations around the world. Finally, this research might be collated using software of the kind being developed at Infosys. In this way, it might be possible to develop a drug, fast, that would be both effective against the disease and useful for the maximum number of people.

The key to systems biology is amassing a wide range of data, and lots of it. 'Our website probably contains the largest data on tuberculosis,' says Thomas. 'We have people who do chemical synthesis, basic biology, genomics, proteomics, biotechnology, high-performance computing, high-throughput, all of this.'

'I think OSDD is a great project,' adds Kitano. 'They are not only scientifically motivated, but also emotionally motivated, as tuberculosis is their problem that has been ignored by Western pharmaceutical companies. This makes members of the project very united and driving the project.'

Modern researchers, he says, can learn a lot from this Asian-inspired way of carrying out research because it 'heavily depends on experience and case-based accumulation of practices.' OSDD's advantage is that, using systems biology, it attacks the problem from all sides. In fact when Kitano describes systems biology in scientific papers, he uses the circular *yin* and *yang* symbol from Chinese philosophy to illustrate how different branches of science – genomics, computation, technology and analysis – can all feed into each other.

The difference with this way of doing medical research is that it involves computer nerds just as much as laboratory scientists. Luckily, India has plenty of both.

'We do have the talent,' says Sulagna Banerjee at Anna University in Chennai.

'The major issue for me is I don't have a lot of things here in my lab,' she continues. 'But I just let OSDD know I need a mass spectrometer and they say they'll find someone from somewhere, either in India or abroad. Speed is the major issue. You can use a global resource. You're not stuck. The intercommunication is amazing. The OSDD project has got us together, and we find we speak the same language. So far, just look at their website, they've thought of almost everything.'

The geeky manpower that is supposed to underpin the project was tested last autumn, when the OSDD team launched a subproject to annotate the entire tuberculosis bacterium genome. The way this annotation works is that a researcher will compare sections of the tuberculosis genome with sections in similar organisms, and then draw conclusions about what these parts of the genome represent. It's exactly the same process as comparing the zebrafish genome to the human one.

This kind of work normally takes a small team many years, but Zakir Thomas believed that with enough people involved online, they could do it in far less. They called the project Connect to Decode (decoding is another term for annotating genomes).

Banerjee was among the members who roped in some of her research students to be part of the project. 'They invited over eight hundred participants from all over the country, and registered them into this common platform,' she says. 'Initially I had a hundred students who were interested in this and then, depending on their ability and interest, the number came down. I think forty students went through the entire thing. They're from all over the country, from Delhi, Rajasthan, Punjab, Kolkata, and also from this area.' When the OSDD team recently organised a conference for the students involved in Connect to Decode, more than 500 turned up.

With Banerjee's help, the group finished its section of the genome far quicker than she had expected. 'Within two months we had finished. It was an immense workforce. We really couldn't believe that we got it done in such little time. Right now we're just compiling it, looking for mistakes and typos and stuff so we can finally upload it to the site,' she says.

But there's still some doubt about the effectiveness of this approach. Scientific journal articles have questioned whether open source can ever really work in biology, where research is so disparate and requires years of expertise, not herds of nerdy kids in front of computers. A recent article in the scientific journal *Nature* complained that the OSDD team still haven't published its research, so fellow scientists haven't yet been able to check it.

Banerjee admits that usually you need specialists to do this kind of work. But she insists that, with the plentiful supervision and training they received, the Indian students have met the standards and their work will be published soon. 'I was so amazed that these are masters, even bachelor-level students, that were getting back to me with such questions. I was really impressed. I did not expect masters students to be asking such questions at that level.'

'So you think OSDD can work?' I ask her.

'So far everything's been taken care of,' she says. 'We just have to see whether it works or not. Just for the science part of it, this is going to work. This is the future. You just have to believe in it.'

I spend that evening in my gloomy hotel room, dining on a damp room-service *dosa* filled with potatoes and watching old-style south Indian soap operas.

It's easy to get swept away by new scientific ideas, but I have little sense of what the rest of the scientific community thinks about this different approach to drug discovery. Do they even take it seriously? To get an outsider's perspective, I call up a British scientist I know, who has also been researching the tuberculosis bacterium.

'You get people really guarded here. Seriously, you can get really despondent in academia in England,' she says, joking that some people go so far as to hide their notebooks from one another, even in their own research groups. 'It's one reason why progress can be slow. So open source will be a big thing,' she adds optimistically.

She tells me that she once visited a pharmaceutical company in Bengaluru. 'India just seems different. People I spoke to over there were just really keen to work together. They weren't precious about it.'

India's edge isn't so much in its research capabilities, which still lag behind those in the US and parts of Europe, admits Zakir Thomas back in Delhi, but in its vast, geeky people power. 'No one else in the world could have taken up this challenge and said in three months time we will completely annotate the tuberculosis genome. Why? Because we have the people,' he says. In this country, uniquely, there are thousands of students who are willing to spend their time painstakingly picking through the millions of letters in a long string of DNA. I wonder why this is. Perhaps it's out of necessity – a functional need arising from the fact that there simply aren't enough scientific resources to go around, forcing researchers to collaborate.

But it's not just about ownership or manpower, Thomas tells me. Science here isn't as constrained as it is in more regulated and established Scientific Societies, such as those in the West. He believes that this freedom is what gives Indians the space to try something different and take risks.

I admire his faith. It reminds me of that story of the mythical elixir of immortality, distilled from the oceans. The irreverent masses of Generation Y will change the way science is done, I was told at the start of my journey, and here near the end of my geeky trip, I'm finally starting to see how that might be true.

Yet despite the endless optimism I've encountered so far in the medical laboratories, I can't escape the fact that the tuberculosis drug hasn't been invented.

My final question for Zakir Thomas is if he truly believes that OSDD will yield a new medicine. He pauses, looking at me with the hint of desperation that I've seen so many times in India, sometimes in a begging stranger, sometimes in the last-shot salesman. I start to wonder whether this hopeful fear is really what drives everything here, including the science.

'See in science, success is a matter of luck,' he says, slowly. 'Failure is a matter of right. Unless we deliver on the ground, people won't believe in this model. Will we get to where the whole world's pharmaceutical industry has not succeeded in the past thirty, forty years? It's difficult to say. But just yesterday I was in Bengaluru talking to a pharmaceutical company about a promising lead of one of our scientists. So there is a possibility.'

'So you think the chances are good?'

'Our chances are better.'

CHAPTER 8

BRAINPOWER

There's one event that I can't miss. It's the annual Indian Science Congress.

As far as scientific conferences go, it has to be the wackiest. It's more like a *Star Trek* convention or the International Comic-Con. First of all, there's no bar on what kind of research you do, so you'll find agriculturalists, software engineers and rocket scientists all on the same bill. This indiscriminate cross-section of Indian science extends to the bookcases in the exhibition hall, weighed down with indecipherable tracts on mathematics and stacks of dubious pamphlets on the science of spirituality.

Secondly, there's no cap on how many people can come. So the website is jammed with enquiries (they're expecting 5,000 delegates and 3,500 students this year). Nor is there a limit on the number of scientific papers that can be presented (around 2,000 at the last count). The list of lecture topics is a cornucopia: It includes 'The ornamental fish trade in the Red Sea and the Gulf of Aden', 'The draught ability of bullocks' and equally oddly spelled, 'Menopause: the end of reproductive phage of the women's life'.

The 97th Indian Science Congress will be the biggest yet. It starts just after New Year, with the pop of fireworks still echoing, when the temperature dips just enough to allow hordes of scientists and engineers to squeeze into sweaty lecture rooms and fight over the last plastic cups of tepid water at teatime. The location of the nerdy extravaganza is the University of Kerala in the southern state's capital city, Thiruvananthapuram (the same place that is home to the Vikram Sarabhai Space Centre, where I started this journey). And the reason for this is that the theme, following the success of the *Chandrayaan-1* moon mission last year, is space.

Since the prime minister usually turns up to make a speech, there's always a buzz around the event, but the excitement seems extra feverish this year. It feels as though India has changed. The software industry is enjoying a boom, statistics show that research output is going up faster than ever, and Indian geeks, rather than emigrating to work in the West, are choosing to stay here instead. So huge though it always is, this year's congress is different. A former president is coming, along with a string of other cabinet politicians and two Nobel Prize winners from overseas.

But what I still want to know is if India has its sights set beyond the changes that have already happened, if it takes seriously its ambition to become a scientific superpower. Britain, Germany, the United States and Japan are the countries that, over the nineteenth and twentieth centuries, were responsible for the greatest leaps in theoretical research and the most valuable inventions. Is India in their league?

My tickets are booked.

Before I leave on the last long flight back to the south, though, I have one more place to visit. It's called Trombay, a suburb just outside Mumbai. And this detour is as much a personal trip as it is an important part of understanding the future of this nation of geeks.

In 1968 my dad, who had just finished a degree in chemical engineering in the north, went to the Bhabha Atomic Research Centre in Trombay for a job interview. Jawaharlal Nehru had died four years earlier. It would be a few more decades before the start of the technology boom, and unemployment among scientists and engineers was high. This research centre was one of the few places hiring engineers and my dad didn't get the job. Instead he went to work for an industrial gas company in Kolkata. Eventually he joined thousands of other Indians who emigrated that decade, seeking out better opportunities to stretch their brains in the West.

And that's how I ended up being born a British geek instead of an Indian one.

More than forty years have passed. The world is different. There are 5,000 scientists and engineers working at the Bhabha Atomic Research Centre. Unlike in the 1960s, India now has nuclear weapons and a resurgent nuclear power programme – most of it devised here. The facility is so secure that it takes three months of emails, phone calls, being shunted from official to official, before I finally receive unexpected notice that I'm allowed inside. Not even all the way inside, they tell me, just as far as the training school, which is about a kilometre and a half outside the inner sanctum, where they run their top-secret experiments.

If there's one thing that definitively separates industrialised countries from poor ones, it's how much energy they use. And this centre was designed to power India's growth after independence by building a generation of mighty nuclear power plants. Unfortunately it turned out to be a more expensive and complicated plan than first envisaged by leaders like Nehru. Uranium – the most popular nuclear fuel – is difficult to obtain. There are still only nineteen nuclear reactors in the country, providing around three per cent of the electricity mix.

So the country still relies on its old coal, gas and hydroelectric power stations. To this day, they struggle to meet its electricity needs. Around two in five Indians, most of them living in rural areas, have little or no access at all. Businesses in the major cities often have backup generators to compensate for the urban power cuts. And the problem gets worse every day.

But the Bhabha Atomic Research Centre is busy once more, this time bridging the growing energy gap that is being created as the country steams into the technological age. The big difference this time, though, is that the nuclear scientists here are inventing a completely new kind of reactor. It promises to be cleaner, safer and more efficient, and provide an almost endless source of power to India.

The centre is a gigantic, Soviet-style complex of buildings, spanning eight kilometres. Surveying the décor, I imagine not much of it has altered since my dad's time. A television screen on the brown wall in the 1960s-style corridor shows a live bar chart of the radiation levels at the nearest nuclear sites in India. There's a sign on the noticeboard telling the trainees who are staying in the dorms that there's been a water leak, so the taps don't work. Water is instead being brought in every day by bucket.

The physicists I'm supposed to be meeting are late and I've been here for more than half an hour already, slowly sinking into a soft beige-leather couch in the waiting room. I'm handed a leaflet to read. This vast facility was named after Homi Jehangir Bhabha, it says, who was a prominent physicist and one of Nehru's close scientific allies. On the inside cover is a small portrait, revealing that he had a dimpled chin and dark, almond-shaped eyes. He looks almost like a Bollywood movie star. Underneath the picture is a quote, dated 1955:

'For the full industrialisation of under-developed countries, for the continuation of our civilisation, atomic energy is not merely an aid.

It is an absolute necessity.'

'We'll eat first, then talk, OK?' says a tall man, walking into the waiting room. Mr Sharma works here as a liaison between the nuclear scientists and the public, he tells me, and he's escorting me to the canteen for lunch.

There is rice, salad, vegetable curry and *roti* in large containers, followed by thick slabs of ice cream for dessert. Three researchers and I squeeze next to one another on low seats, our plates in our laps, awkwardly attempting to strike up some casual conversation. One of them tells me excitedly that all the kitchen scraps and leftovers here are converted to cooking gas, using microbes that feed on the material, so that it doesn't go to waste. And a group of researchers here, another says, have also invented a radiotherapy unit for cancer patients. They call it the Bhabhatron. The scientists around me smile and nod.

I suspect they're nervous that I might be anti-nuclear.

After the famous nuclear disaster at the Chernobyl power plant in Ukraine, in 1986, which killed at least 4,000 people, the industry suffered an irreversible knock to its reputation. Nuclear power became seen as an environmental villain. By the end of that decade, nuclear reactors around the world were gradually allowed to reach the end of their working lives without being replaced. Nuclear scientists became a dying breed (literally: universities stopped teaching the subject and former experts died of old age).

I've been to two nuclear power stations before. The first was Sizewell B in England, a beautiful white-domed building by the sea. And the second was a Soviet-era plant in a silent corner of Lithuania called Ignalina, built to a similar blueprint to the one in Chernobyl. One thing I've noticed about both of them is that, just as here, the workers are evangelically pro-nuclear. Not because their employers brainwash them, but because working in a place like this every day, for years, teaches them that nuclear reactors are far safer than the outside world realises. This is one of the most heavily regulated industries in the world. And contrary to the myths, radioactive leaks are almost unheard of.

But the past has left its mark, and so every remaining power station feels somehow old. They're like relics of a science-fiction age, not long after the early days of atomic physics, when people first believed that uranium would solve the world's energy problems.

That age began on 16 July 1945, when the world's first atom bomb sent a mushroom cloud into the sky over the deserts of Jornada del Muerto in New Mexico. Robert Oppenheimer, the father of the Manhattan Project – America's attempt to build the world's first nuclear weapon – was on the ground. He's reported to have said afterwards that a few in the small crowd around him laughed, while others cried. But, as a sometime student of Hinduism, he instead happened to remember a line from the ancient Indian religious epic the *Bhagavad Gita*.

In one scene in this long poem, there's a battle for supremacy between two powerful armies. But a warrior prince loses his nerve and refuses to fight. So to urge him on, the god Krishna transforms into Shiva, one of Hinduism's most feared deities. Blue-skinned and four-armed, there are majestic statues of Shiva sitting crosslegged, garlanded with a snake and brandishing a trident, dotting ancient temples from high up in mountainous Nepal to down south in Thailand. Mighty and devastating, Shiva awes the prince into fulfilling his duty in battle.

So when he saw the mushroom cloud, Oppenheimer quoted a line attributed to Shiva at the moment of his transformation. 'Now I am become death, destroyer of worlds.'

Part of the destructive beauty of nuclear power is that it's possible to get vast amounts of energy from relatively tiny amounts of fuel. Enormously strong forces hold subatomic particles together inside an atom, and if one of these particles escapes, the energy is spontaneously released in a burst of heat and light. A few materials, including uranium and plutonium, emit these particles over time naturally, in a process known as radioactive decay. When they decay, they release this energy, which is why radioactive things sometimes give off a warm glow.

But the process is too slow to generate power on a large scale. It takes more than four billion years, for example, for half of the atoms in a block of uranium to decay. So scientists can instead artificially release this latent atomic energy by bombarding an atom with subatomic particles, forcing it to split. Breaking a single uranium atom releases 200 million electron volts. A big number though this may seem, electron volts are actually minuscule (200 million of them are too tiny to light even a small bulb). But when one atom is broken, more subatomic particles are released, leading to a chain reaction as they hit other atoms. Since a kilogram of natural uranium contains more than two trillion trillion atoms, this means that just one kilogram of uranium fuel in a nuclear reactor can produce an enormous amount of energy: the equivalent of burning 16,000 kilograms of coal.

The challenge for physicists and engineers is controlling this nuclear reaction and maintaining it in a state of balance, so it produces useful power but doesn't run away into an explosion. Inside a reactor, nuclear processes have to be self-limiting. If the temperature goes up, for example, internal mechanisms should naturally slow down the nuclear reaction so that less heat is produced and it returns to equilibrium. And when everything is under control, the reactor needs to operate continuously, releasing a steady flow of heat. If it all works, this heat can turn a channel of water into steam, which can be used to turn turbines, which produce electricity at the other end.

US scientists began building their first full-scale atomic power station in 1954 in Shippingport in Pennsylvania, to try to prove that they could produce enough electricity for 100,000 people. The small-scale experiment eventually wound up in 1988, but it was enough to demonstrate that nuclear power was technically possible. By 2005 there were 443 power stations in thirty-one countries around the world, mainly in the US, France and Japan.

But the most interesting thing about the Shippingport reactor, other than being America's first, was that it used a different fuel from nearly every power station that followed it. Most plants nowadays use a type of uranium called uranium-235. In Shippingport, they used thorium.

Thorium is a lesser-known radioactive metal, tucked away in the bottom left-hand corner of the periodic table of elements. To look at, it's a soft, silvery-coloured substance that turns grey and finally black if it's left out in the air for too long. And when heated, it burns with a brilliant white light. Thorium, like uranium, exists naturally in soil and mineral sands, but is at least three times more common than uranium and easier to mine.

Thorium's abundance isn't the only characteristic that made it an attractive fuel to the engineers in Shippingport. It also absorbs particles more easily than does uranium, which means that more energy is released by thorium than by the same amount of uranium. This also means that it creates less waste. Because of its particular place in the periodic table, the waste that does come out of a thorium reactor also contains a tinier proportion of really long-lived waste elements. And this in turn means that it doesn't need to be stored under such tight conditions or for so long. While uranium reactor waste has to be buried for hundreds of millennia, thorium is relatively safe after just 300 years. Altogether, this makes it a plentiful, relatively cleaner and arguably safer source of nuclear power.

At the time, though, thorium was dismissed in favour of uranium. As far as modern science has established, it's impossible to make a bomb with thorium, which – in the 1950s – was not a plus. Since uranium fuel can be refined and turned into the explosive kernel of a nuclear weapon, it killed two birds with one stone. It gave nations like the US and the Soviet Union nuclear power while at the same time arming them with atom bombs. So in the decades after the Shippingport reactor was built, uranium became the nuclear fuel of choice around the world. By the early 1970s, new studies into thorium began to dry up. And since the 1980s, there hasn't been another thorium plant producing electricity.

Little did anyone know that, fifty years later, thorium would start looking good again.

The fear of nuclear apocalypse has been replaced by the fear of climate change. Fossil fuels like coal, oil and natural gas – already becoming scarcer and pricier in many parts of the world – are now known to cause global warming too. The bonus of nuclear power is that it's pretty much carbon-free and so, in the ongoing search for alternative, cleaner energy sources, new nuclear power stations are being built and radioactivity is enjoying a renaissance. Ironically, given its reputation since Chernobyl, nuclear power is now being seen as an environmentally friendly fuel. The World Nuclear Association predicts that by the year 2100, there could be at least 2,000 reactors around the world.

The problem this time is that stocks of uranium are limited, and there may not be enough of the precious fuel in the world to support such rapid growth. So now scientists in countries including France, Japan and Canada are revisiting the thorium research that has been neglected for so long.

And in a historical twist that even Oppenheimer himself couldn't have predicted, the one nation that's poised to build a new generation of large-scale, modern, thorium-powered nuclear reactors before anyone else isn't the US, Russia or a country in Europe.

It's India.

'In India the supply of thorium is at least eight times that of uranium,' says Dr Ratan Kumar Sinha, the 59-year-old director of the reactor design and development group at the Bhabha Atomic Research Centre. He has a small round face, a little grey moustache and thick glasses covering his bug-like eyes. There are two other scientists in the room with us, both engineers working under Sinha, and both endearingly shy and polite. They let their boss do the talking. He talks fast. 'We have so much thorium,' he continues. 'It's lying on the beach. It's on the sand for all to see and it is very good quality. We expect the reserves to be the size of 800,000 tonnes.'

Backed by the government, Sinha heads the group that is creating the world's next generation of thorium-fuelled reactors. It is research that has been happening on this site since 1955, in anticipation of India one day needing it.

The biggest problem since India started its nuclear programme has been the fine line separating research into nuclear power from work on atomic weapons. In 1974 scientists at the Bhabha Atomic Research Centre detonated India's first atom bomb in an underground explosion in the desert city of Pokhran in Rajasthan. The country immediately became a scientific pariah; Europe and the US severed many of their science and technology links with India for fear they might abuse information and use it for defence research. But more devastating for civilian Indian scientists such as Sinha was that they were cut off from experts elsewhere in the world.

Since that decade, both Indian nuclear geeks and space geeks – since constructing space rockets is only a short leap away from building missiles – have been forced to be self-sufficient, and it has slowed down progress. Research and engineering has to be done almost entirely from scratch and independently.

But then this also means that scientists in India tend to think outside the box.

'Currently the growth of India is about eight per cent per year,' says Sinha. 'Six to eight per cent.' He has a knack for carrying statistics in his head. 'Now that requires energy supply to increase by nine to eleven per cent. And if you extrapolate that, it leads to very large requirements for the future. At the moment there is a supply and demand gap, which is why there are power cuts in many places in the country, even today.'

Sinha passes me some documents to read. Among them is a presentation that the Indian delegation made to the International Atomic Energy Agency a couple of years ago, spelling out the benefits of thorium fuel. 'POWER POTENTIAL IS VERY LARGE' says one line.

He begins by explaining just how bad India's electricity shortage is. 'There are two or three parameters which combine together to give a human development index,' he says. 'It's a measure of developing capability in a country. And a figure of 0.9 and above is considered to be good. Most developed countries lie within it, taking 0.85 as a transition point. And it turns out there is a good correlation between the human development index and the electricity available. It says you require to have something like 5,000 units of electricity available per person per year to be able to support the educational and healthcare and industrial infrastructure for quality life to be developed today.'

I do a quick calculation in my head. This means that a country the size of India would need at least 5,650 billion watts of electricity every year to give its citizens decent living standards.

'How much is the country producing now?' I ask.

'At the moment we're nearly 150 billion watts. So it's quite low,' he says. 'We need to go further, and if you wish to go to that level, even to 1,200, and use only coal, it turns out that we would end up burning coal at the rate of almost four and half times that of the United States today. In fact we don't have that good enough quality of coal anyway. We would have to import coal.'

The other concern is that, under international environmental commitments to lower carbon emissions, India has to cut back on fossil fuels like coal. On top of this, because of dwindling uranium supplies, if it were to transfer power production to ordinary nuclear plants, it faces the risk of running short of fuel. The government wants to raise the proportion of electricity derived from nuclear power to ten per cent by 2035, but in the last year some nuclear reactors have already been forced to operate at below optimum capacity because uranium stocks have run so low.

In principle at least, thorium could fill this gap. One thorium fuel pellet can light a bulb for 200 years, says Sinha.

'The regulators demand that any new design, first-of-a-kind, has to be supported by a lot of experimental data,' he explains. But the reactor design team here at the Bhabha Atomic Research Centre has proved that it's theoretically possible. 'The second part is the physics,' he says. Tea arrives in small cups, with biscuits. We pause while Sinha takes a sip.

Inside this vast campus is a critical facility, which they're using to test out their theories. 'It has to be based on calculations, taking into account scenarios that engineers will pose. We have started with uranium-based fuel. In a little time, in a few weeks, we will start putting thorium in there,' he continues.

'Are you confident that it will work?' I ask.

'There is no doubt.' In fact, Sinha envisages that nuclear power will ultimately provide half of India's electricity needs. 'We're considered pretty advanced in this technology in the world, you know,' he adds.

'First American reactor was thorium. It was in Shippingport,' says Dr Viswanathan Jagannathan, one of the other two physicists in the room. 'But that was only a demonstration. Those early reactors were to make a point. This is not to make a point, it's an engineering solution.'

One of the hurdles to building a modern, large-scale working thorium reactor is that, unlike uranium, thorium can't start a nuclear reaction on its own. Instead, the metal has to be bombarded with subatomic particles until it decays into a completely different material, another form of uranium, known as uranium-233. This is a little lighter than uranium-235, which is the type normally used in power plants, and makes a more efficient fuel.

Since 1996 the team has been operating an experimental reactor in the township of Kalpakkam, near Chennai. It's the only reactor in the world to run on uranium-233, Jagannathan tells me. And it's part of the proof that India needs to convince the world that its large-scale thorium power plants can work. 'There are some reactors using thorium on an experimental basis, but very few,' adds Dr Pritam Das Krishnani, the other physicist in the room, in a whisper. I get the sense that he doesn't get out of the laboratory much.

By any measure, India is leading the way in thorium research. Scientists here publish more journal articles on the subject than in any other country. And since the success of the experimental model in Kalpakkam, the team has started working on a real thorium reactor. Jagannathan pulls a yellowing scientific article out of a file to show me how it works (he can't show me recent documents because they might give away research secrets).

There's a drawing of the inside of a thorium reactor. To kickstart the reaction, he explains, the engineers mix the thorium with a smaller amount of uranium. The picture illustrates a cross-section of how the nuclear fuel rods look: eighty-four of them stacked in a circular cluster and represented by different colours depending on what kind of metal they contain. The red ones, which hold uranium, are on the inside of the reactor, while the green ones, which carry thorium, are on the outside. There's a circular hunk of beryllium metal in the middle, which produces some of the subatomic particles that will smash into the atoms and help break them apart. The whole thing is packed tightly inside a pressurised tube.

The powerful nuclear chain reaction inside this tube is kept under control by a blanket of heavy water. This is like ordinary water, except with atoms that have extra subatomic particles inside, which both slow down the subatomic particles shooting around inside the reactor and mop up any stray particles.

From the date on the article I can see that Jagannathan wrote this paper more than a decade ago. And the research has moved a long way since then. The details are classified but, he tells me with a grin, the team has already designed a commercial mixed-fuel mini-reactor, in which about 39 per cent of the energy comes from thorium, with the remainder from uranium. They've named it an Advanced Heavy Water Reactor.

It's designed to be so safe that even deliberate neglect and sabotage can't drive it into an explosion. 'From outside, if somebody chops off all the electrical lines and you have no diesel generators, if someone chops off all the sources of cooling and there's no seawater, and if there's no control room or operators,' says Sinha. 'If you conceive of this scenario, this reactor takes care of itself.'

I can glean the basics from Jagannathan's diagram. Like a lot of engineering structures, the safety lies in the symmetry. A nuclear reactor needs to be geometrically precise because the pressure balance can be messed up if just one rod containing the nuclear fuel is slightly askew or the wrong size. There are 360 fuel rod clusters, with ninety-one extra thorium rods, all arranged in a perfect hexagon. There are pink rods, which are the final ones to be used in the fuel cycle, bordering the hexagon. At the centre are white rods, used second to last, arranged in a Star of David, resembling ancient artwork that I've seen on the walls of Mughal monuments in New Delhi. But this is just the core. Outside this, there must be layers and layers of protection.

'This Advanced Heavy Water Reactor is for export market, for short term,' says Sinha. Their creation is so groundbreaking that they're planning to sell it to the rest of the world, as a potential solution to the energy problems faced by smaller countries that are trying to reduce their reliance on fossil fuels. It has a working life of a hundred years, perfect for nations that need new power sources soon, he adds. According to the brochure, they expect it to be on the market by 2020.

India's domestic thorium reactors, the trio tell me, will be even bigger than this. They talk about the plans with such nonchalance that I almost forget that they have the nation's future in their heads, ready to be turned into blueprints that will map out the infrastructure that could power millions of homes, schools and perhaps – if the correlation between electricity usage and standards of living is correct – finally tip the population out of poverty. Homi Bhabha's old nuclear facility could finally live up to the promises it made back when my dad was here.

The tea is finished and my time is up. They all have to get back to work in their laboratories.

And I'm heading to the 97th Indian Science Congress.

ROCKET SCIENCE

The lecture room is a steamy reminder of just how desperately the University of Kerala needs air conditioning. Rows of scientists fan themselves with magazines, shifting uncomfortably in their seats. I spot a lazy professor ordering his assistant to fetch him a cup of water. Even the bugs are too tired to bite. It may be the height of winter but here, just slightly north of the equator, it's hot; too hot really for a conference. My skin reddens and leaks fat beads of sweat while I walk from this building to the next one, which is so frustratingly distant that I'm tempted to skip everything and take a nap.

The crowds get bigger and a pair of ambulances arrives. 'It's a few children over there,' someone says. Local schoolkids have become dehydrated while waiting to get into a science exhibition and a few of them have fainted.

For those of us who manage to stick it out in the stifling lecture halls, stretching our necks towards the ceiling fans, the Indian Science Congress is a mixed bag.

Among the better debates is one on India's energy needs, packed full of Indian power experts. 'Coal may not last more than eleven or twelve years . . . uranium will last only a few years,' argues the former chairman of the Atomic Energy Commission, Anil Kakodkar. 'Under this scenario, it looks to me, as far as indigenous resources are concerned, India will have to rely on two things. One is solar. The other is thorium.'

Two smiling women in white cotton saris – one of them looks as though she may be a European hippy – turn up near the end of the debate, climbing onto the platform with the speakers. Handing out thin orange leaflets with a baby's face beaming from the covers, they tell the crowd there's another form of energy that's even more important than heat, light and electricity. According to the handouts, they're from a group called the Global Energy Parliament, which seems to be part of some local spiritualist movement. 'The thought vibrations generated by human beings has [sic] not been properly studied,' it reads. 'Unfortunately our human vision of energy is circumscribed by the boundaries set by scientific assumptions of the early 19th century.'

'You never know, they might turn out to be right!' says an elderly man sitting next to me, laughing.

People accept the leaflets politely, dropping them in their bags, unlikely to be read. We shuffle outside, ambling towards the refreshments tent to stand in the queue for snacks. I've forgotten my meal vouchers so I sneak in round the back, dodging a pair of security guards and attracting the disapproving stares of onlookers. The tiny paper plates of doughnut-shaped *vadas* aren't worth this effort, but I have no choice, given that the nearest restaurant is at least a mile away.

Meanwhile hundreds of delegates sneak away, disappearing on air-conditioned coach trips to nearby tourist sights. The favourite destination is Kanyakumari, a legendary spot on the southernmost tip of India at the meeting point of the Arabian Sea, the Gulf of Mannar and the Indian Ocean. Here the Hindu monkey god Hanuman was believed to have dropped on the ground a chunk of mountain containing lifesaving herbs. It's the reason for the area's lush biodiversity, some people say.

Although it doesn't seem that way from inside this university campus Kerala is a tourist magnet. From the air, as my flight came in to land, it reminded me of a tropical jungle I saw once in a movie about the Vietnam war. The backwaters look like thick pools of brown soup, swirling with neon-green algae and garnished with palm trees. From the ground, it's a sleepy city riddled with enormous mosquitoes, gorging on the sunburnt flesh of Westerners. This southern corner of India has been dubbed 'God's own country' to attract holidaymakers. Tables outside the beachfront shops display fresh lobsters, butterfly prawns and white fish split open on melting ice, their entrails dripping onto the sand. Hordes of Europeans buy overpriced turquoise beads and sip on tall glasses of lime soda, while blue-sailed fishing boats float past.

But the centre of the capital city, Thiruvananthapuram, is very different. There are high-rise office buildings and red posters plastered all over the streets, showing fists held high in the air and images of Lenin, Marx and Che Guevara. Despite the rampant commercialism and Western invasion on the coastal resorts, in reality, like Vietnam, the Communist Party (democratically elected here) still controls Kerala.

And this has given the state a unique social structure compared to the rest of the country. Across Kerala, I'm told, literacy rates are high (in fact, this was the first Indian state to achieve one hundred per cent literacy). Many people speak English as well as the local language, Malayalam. And at the congress, female scientists appear to be almost as common as male ones.

Keralites aren't just clever, one of the delegates informs me, they're adventurous too. 'There are Keralites everywhere – Dubai, Canada, the US, UK,' he says. 'The joke here is that when Neil Armstrong landed on the moon, a Malayali was already there, offering him a coffee!'

Perhaps because of the tourist industry, perhaps because of the communists, the auto-rickshaws and taxis can be expensive. The one I finally caught to get here, after realising that I wasn't going to find a better price, had a kitsch picture pasted next to the passenger's seat of a fat, smiling baby, sitting in a mosque and reading the Qur'an. On the way from the airport to the Kariavattom campus of the University of Kerala, the scenery changed again, the coconut groves giving way to swanky high-rise apartment blocks. One of these new housing developments is called Cyber Palms, and another, still only a mess of rickety scaffolding, had already been named Cyber Heights.

Part of the university lies inside the city's technopark, built fifteen years ago to attract software firms from booming Chennai and Bengalaru to this quiet, geeky city in the south. Gaslight-style lamps dot the spotless, tree-lined roads, with small office blocks nestled in between the hilly curves. The entrance to the campus is a mile or two past some iron gates, which are patrolled by police officers – the prime minister is due to arrive soon, and security is tight. There's a branch of the American fast-food chain Subway just inside the entrance and a trendy coffee shop next door to that.

So now, at the overcrowded refreshments counter, unless I want to brave another long and expensive rickshaw ride, it looks as though I have no choice but to eat what I'm given. I stand in a corner, picking at my plate with a plastic spoon, while someone spills tea on my sandals.

After lunch, I wander to a session on the medicinal plants used by Indian tribal communities. It's in a dark first-floor room with orange curtains and green metal seats, containing around twenty or so people, almost all of them old. Flicking up pictures of herbs on the screen behind him, a researcher claims that the plant *Angiopteri Erelta* is a useful remedy for 'leprosy, ulcer and cancer'. Yes, all three, he insists when I ask whether this is in fact true. The only problem, he admits, is that these claims haven't been scientifically tested. A handful of lectures like this one, relegated to the far, shadowy corners of the campus, seem to sail dangerously close to the wacky.

The next speaker tells us he's found a new cure for avian flu. 'The size of the virus bears no relation to the havoc it has streaked upon human health,' he says. Does he mean streaked or wreaked, I wonder to myself. But the dictionary abuse is replicated on the slide in front of us. 'The world community has spent billions of dollars in fighting this thread,' it says.

Finally, just as I'm ready to leave, he reveals his miracle remedy. The avian flu epidemic could have been prevented, he says, by using a homeopathic treatment called *Pulsatilla Nigricans*. The talkative moderator, an elderly man with yellow buck teeth, agrees, closing the strange session with a lengthy tirade against modern medicine. 'There are 15,000 medical journals in the world but after fifteen years, each one becomes null and void,' he says, matter-of-factly.

I go back outside, passing busier lecture rooms containing young medical students. Anywhere else in the world, the presence of spiritualists and herbalists alongside genuine researchers might be considered an affront to modern science, but what surprises me about the Indian Science Congress is that the delegates seem peculiarly happy to give them space to air their views. And given that all societies have their oddballs, I think to myself in a haze of sunstroke, where's the harm?

At the start of my trip, I might have found this kind of communion between the scientific and pseudoscientific unfathomable, but now I can't help thinking that India's precarious place in the scientific landscape, between the geeky and the bizarre, may not be an entirely bad thing after all. Maybe it's just the effect on my brain of the blazing sun and the dry snacks, but I'm starting to think that the fact that even the wackiest ideas are tolerated in India has given scientists and engineers here a unique freedom to explore the edges of what's believed to be possible. Without this, perhaps biologists would never have attempted open source drug discovery to find a new cure for tuberculosis; perhaps Bengaluru's geeks would never have built their global software companies; and maybe thorium reactors and manned space rockets would still be trapped in someone's wild imagination. Those on the scientific fringes – whether they're disastrously wrong or gloriously right – are not simply the freak shows at the fair. In India, they're the main event.

It's an aspect of scientific discovery that seems to have long been forgotten in the West, where I grew up, where science is in danger of being thought so boring and straitjacketed that children no longer dream of becoming engineers, inventors and laboratory researchers, as they do here.

Next on my schedule is one of the most well attended talks at the congress, being given by a 75-year-old chemist from Bengaluru, Chintamani Nagesa Ramachandra Rao. Celebrated for his work on the structures of materials, today he is the only person to be a member or a fellow of all the major scientific academies of the world, including the reputed birthplace of modern science, the Royal Society in London. Rao isn't just a geek, he's a nutty professor. He looks like one too, with a giant mess of wiry, grey hair and oversized glasses.

The theme of his lecture is how to encourage scientific curiosity. He admits that being an Indian researcher has never been easy. Science isn't really about the big projects, he says, it's about the thousands of ordinary researchers in their laboratories every day, making incremental steps in human understanding. And unfortunately for them, India's smaller laboratories are still poorly funded and university departments often feel weighed down by hierarchies and frustrating bureaucracy. He hopes this is changing.

'I really want India to become great in science. All other things will shine when science shines,' he says, to thrilled applause from young, starstruck students in the front row. Good science takes passion, he tells us. 'We need real nutty guys doing science!'

By the next day, I can barely move for the crowds.

Today has been given over to the focus of the congress – Space. There's something especially fascinating about space research. The heavens have always obsessed the most curious minds, and perhaps that's the reason why so many of the world's earliest scientists were astronomers. The challenge of that final frontier is what pushed the US and the Soviet Union into a space race in the 1960s. And even now, having a presence outside the Earth's atmosphere has become the threshold between the most scientifically advanced nations and the rest.

Here at the congress, it feels like an obsession. There are packed seminars on satellites, mining space resources and extraplanetary travel. Gossiping researchers speculate on how far India will reach in its ambitions. Can it really achieve what only the US, Russia and China have, and send people into space?

In a quiet gap between the lectures and seminars, I find myself sitting next to Narayana Moorthy, a researcher at the nearby Vikram Sarabhai Space Centre. He's a short, unassuming man, local to Thiruvananthapuram. And, I learn, he's the person responsible for building India's first manned rocket.

Like nuclear power stations, the Indian space programme has been dogged by the shadow of the Pokhran atomic weapons tests, decades ago. Teams of physicists and engineers have painstakingly caught up with the world's other space agencies by intensively building their designs from scratch. And Moorthy's work has been done with barely any outside help. In fact, he whispers, he still has trouble getting a visa to travel to the US because (he believes) they fear that he might come back with their space secrets.

One of the main technical challenges, he tells me, is scaling up: the bigger and heavier a projectile becomes, the harder it is to propel it into space. If India is to have any hope of launching larger satellites and eventually astronauts, it needs to conquer this problem, he explains. This is his job.

Within a few minutes, Moorthy dives into the nerdy technical details. The largest rocket that India has at the moment, he tells me, is the Geosynchronous Satellite Launch Vehicle II, which is forty-nine metres high and nearly three metres thick. It can carry 2,000 kilograms of weight into space. 'Now if you want total self-reliance in the satellite launch vehicle technology,' he says, 'we should be able to launch around a 4,000-kilo class of satellite. So for that we are developing this new rocket called Geosynchronous Satellite Launch Vehicle III.' His team started work on it in 2002, and when it's finished, it will be a monster of a rocket, with a diameter of four metres, carrying up to 5,000 kilograms.

It's only two years away from its first test flight. More importantly, he adds, it will probably one day be used as the launch vehicle for India's first manned mission.

'The entire technology is indigenous. The entire materials, the entire fabrication, the testing, everything,' he says. 'I'm very proud of working in India. Many of us who started working at the Indian Space Research Organisation in early seventies, we're still working there, and we all take pride in bringing up India as a major spacefaring nation.'

The rate at which India has managed to amass space expertise has come as somewhat of a surprise to the rest of the world. But the biggest shock of all has been how it has managed to do everything so cheaply. According to news reports, the country's space budget is around three per cent that of the US space agency NASA. 'When a NASA guy, he came here, he was surprised to see so many facilities,' says Moorthy. 'Amount of money we spent, we told him. He was looking like this.' He stares at me, mouth wide open, before breaking into a laugh.

But then again, India has never been part of the expensive International Space Station and it hasn't sent a person into space. By comparison, neighbouring China, which reportedly has double India's space budget, had its first manned mission in 2003. So despite all the optimism at the conference, I'm unsure that India's nerdy manpower alone can compensate for the cash needed to complete major scientific projects like these. The risks of failure are huge. If a manned mission fails, not only would India's reputation be scarred, but lives could be sacrificed too. Nothing would be more disastrous for the country than to lose its first astronauts to bad engineering.

Moorthy, though, insists that it *can* succeed, and *will*. In fact, he tells me, they even have plans to develop reusable space rockets, which would make regular trips into space both easier and cheaper. 'You can call it something like space shuttle,' he says, pronouncing 'shuttle' as 'shittle', which confuses me for a moment. 'So this being India, it is trying to develop this technology.' A reusable rocket, or shuttle, would make launching satellites far cheaper. At the moment, many space programmes use a two-stage rocket, of which one part gets ditched and the other part returns to Earth. 'But India is trying to make it two-stage, with both stages recoverable. It comes back, and then we recover it from the sea,' he says. It's perhaps one of the biggest technical challenges in the world.

'Now this type of thing has been done is US, in space shuttle,' he explains. 'The motors are recoverable and can be reused. Of course the problem of reusing the already-used hardware is that all this stress is a problem.' Because the components have to withstand such huge forces, they have to be overdesigned to compensate for the fatigue that builds up over the course of every space journey. Even so, Moorthy estimates that this technology, which they've already started working on, will be finished in fifteen years.

Moorthy doesn't even seem fazed by the idea of a manned

mission. 'Definitely it will happen,' he says. 'That's why we have test flights. In fact our confidence has gone up because in 2007 we had a mission called the Spacecraft Recovery Experiment, wherein after twelve days of orbiting a satellite, we recovered it exactly on the spot where we wanted.' He smiles broadly. 'In the Bay of Bengal, we predicted a location and exactly it fell and then we recovered it back. This re-entry technology is very hard, where the temperatures are very large of 2,000 degrees. So this is one of the scientific things we have done – high temperature tiles and that. And experiments, we will have one more in the coming year. This will give us confidence that we can land the man on land or sea or wherever we want. Technologically we can do, that's no issue.

'The only thing,' he adds, 'is we have to create certain infrastructure of testing the humans. Astronautic training centre. And other, this medicine area.' Making a rocket fit for habitation and people fit for zero gravity is the other end of the science of human space travel. In the early 1980s, NASA offered to take up an Indian scientist in one of their space shuttles. Candidates were hastily interviewed, selected and trained. And then a team set to work creating as many as twenty Indian space foods to suit the palates of these astronauts. They included types of rice, a chicken curry and a thick Asian sweet made of carrots, known as *halva*. In the end, the mission was cancelled because of the American Space Shuttle Challenger disaster in 1986, which killed all seven on board and prompted NASA to stop sending foreign astronauts on space missions.

Since then only two Indian citizens have entered space. The first went up in a Soviet rocket in 1984. The other became a naturalised US citizen before she joined the crew of the NASA Space Shuttle Columbia, all of whom were tragically killed when it disintegrated on its way home.

This time, Moorthy says, India intends to be perfectly prepared. Every aspect is being meticulously planned in advance. The most important move came a few years ago, when the government opened its first Institute of Space Science and Technology, here in Thiruvananthapuram. The first batch will graduate in a year. Its students won't just be building rockets, he tells me, India's next astronauts may be among them too.

Moorthy looks up at the stage. The next session is starting soon, and this is my last chance to meet the man of the moment. He's 66-year-old Madhavan Nair, who was until recently the chairman of the Indian Space Research Organisation and is now the president of the Indian Science Congress. He was the scientist in charge of sending the *Chandrayaan-1* rocket to the moon. And years ago he also worked at the Bhabha Atomic Research Centre, making him one of the heirs of Nehru's dream to transform India into a rational, scientific nation of geeks.

I run backstage and identify him by his characteristic halo of jet-black hair. He promises me a few minutes before he goes on stage. 'You want to know about *Chandrayaan-1*?' he starts. He has a quiet voice, almost a drawl. 'I'm happy it turned out to be one of the perfect missions. And beyond that, the scientific data, what we have collected, maybe it will keep our scientists in India busy for the next three years. So it's a very big achievement as far as I'm concerned. It's an investment for the basic science and for the future.'

The next target, he confirms, is to send two or three people in a rocket around the earth and bring them back. That will take six to seven years. After that, they're planning an unmanned mission to Mars. This is the first I've heard of a Mars mission.

Minutes before, I had seen Nair batting away allegations from news reporters that India is in a space race with the rest of the world. And now I'm questioning what the truth really is. There's something unimaginably ambitious about the speed and scale of India's space programme, as if it's no longer content with fulfilling its early goals of sending up satellites so ordinary people could have colour television and cheaper mobile phone connections. Now, it seems as though India has something more to prove.

I wonder, is it a show of strength?

'You see we have to look from the overall context of the global space programme,' Nair replies. 'The major spacefaring nations, whether it is US, or China, or Japan, they have all got ambitions of having the man in space. So naturally the presence of man in the outer space is going to be one of the major requirements for the future space community.'

He pauses, as though he's afraid of giving away state secrets. 'Well, I would say it's a competition among the nations to have supremacy in this area. And I hope that we really . . . India cannot afford to lag behind in that.' It's more about cooperation, he adds, as diplomatically as he had handled the reporters earlier. But then he flashes a quick smile and lets slip a final thought.

'Today we have come a long way,' he adds. 'Today we have established ourselves as a space power.'

I'm in the biggest, frilliest tent I've ever seen. Like Jawaharlal Nehru more than sixty years before him, Prime Minister Manmohan Singh has made the 2,800-kilometre journey from New Delhi to Thiruvananthapuram, just to make clear how important science and technology are to India's future. His visit makes the front pages of the national newspapers. Thousands of people have turned up to hear him. We've been queuing in the sun for hours, we've been frisked and we're still waiting. It's taking a while for him to turn up, and the tent is getting stuffy.

Worst of all, the officials turn off the fans so that we can hear him more clearly when he arrives. This is unfair, I think to myself, since the dais where the dignitaries are sitting is air-conditioned. I must be around a hundred rows from the stage, almost too far to see anything. A nun in a grey habit and black veil shuffles past. A tall policeman in khaki uniform posts himself next to me, almost clipping my ear with his rifle.

And then the loudspeaker bursts into life. A picture of a young boy in a pink shirt flashes onto the giant screens above us. He's on the verge of tears. He has lost his parents in the crowds, we're told. We glance around, a few people calling out his name in case someone might recognise it. Finally, a woman in a lime-green sari runs down the centre aisle of the tent, scooping up the weeping boy in her arms. Everyone cheers.

More waiting. There's a loud crackle and the screens go black. 'The honourable Prime Minister has arrived. All in the hall are requested to take their seats,' says a voice over the loudspeaker. We hush. Manmohan Singh walks in, wearing his signature sky-blue turban. He has a glossy red gown draped over his shoulders, which I suppose is meant to look like university attire, but instead resembles something a champion boxer might wear before entering the ring. Other dignitaries, including the minister of external affairs and the state education minister, are similarly dressed in shiny orange and blue robes, all as if they're ready for a fight.

'We have worked hard to do what is good for science,' says Singh. An economist, he's a famously practical man, who's free of flourish and brief in his words. He talks slowly, listing his priorities for the coming years. 'Ladies and gentlemen, our government has declared 2010 to 2020 as the decade of innovations. We need new solutions in many areas to achieve our goals of inclusive and sustainable growth. In healthcare, energy, urban infrastructure, water, and transportation, to name only a few. We cannot continue with business as usual. Solutions from developed countries are not always applicable. They are often too costly and at times not sustainable.'

Last year, he promised to double India's investment in scientific research and development from less than one per cent of the country's national income to two per cent. But there's a long way to go. If the number of articles published in international scientific journals is used as a measure, India still lags far behind richer countries. In medical research, India is around thirteenth in the world, in mathematics it is twelfth, and in chemistry it's fifth. But then, it is improving this record at a phenomenal rate. While Japan, Germany, Britain and France have become steadily less scientifically productive since 2000, India's research output has risen higher every year. In 2008 scientists and engineers published on average 53 per cent more scientific papers than in each of the five years that went before.

Scientific progress, though, isn't only about the numbers. It's also about nurturing that nutty, indefinable love of answering questions and solving problems. And just as Nehru, Sarabhai and Rao attempted to do when they launched India's space programme, it's also about improving the lives of ordinary people using innovative technology. On that front, while it may not be publishing as many journal articles as other superpowers, India is having an impact far beyond the surface statistics.

I notice newspaper reporters picking up their pens, furiously writing tomorrow's headlines. Singh has announced the launch of the Jawaharlal Nehru National Solar Mission. He aims to build 20,000 megawatts' worth of solar generators by 2020, to complement the new generation of nuclear power stations. It's another grand scientific project to match the space rockets.

He's approaching the end of his speech. 'The time has come to give a new boost to science in India,' he announces. 'I hope each one of you will return to your laboratories and classrooms re-energised to do good science, and do it for the good of our people, and of humanity.' Applause. Singh leaves and we file out of the tent to the busy refreshments counter, technical assistants and rocket scientists alike, blinking in the sunshine.

After a few days, I leave Thiruvananthapuram in an expensive taxi, my books and notes filling a case. Later I spend my flight back to London sorting through this mass of papers. At the bottom I find copies of the speeches that Manmohan Singh gave at the last two Indian Science Congresses. I read them both to the end, trying to pick out any final clues as to where the leader of this nation might be taking it next.

In the last paragraph of both speeches, I notice as the aeroplane jolts off the runway and leaves India behind, Singh borrows some words from the British wartime prime minister Winston Churchill. It's the same quotation each time. Now half a century later when the world has changed, India is free, and Indian brains are set to take over the world, it reads as though it could also have been a script line in a science-fiction movie.

'The empires of the future,' Manmohan Singh told the assembled geeks, 'are going to be the empires of the mind.'

REFERENCES

PREFACE

Author unknown. (Date unknown), *Bakhshali Manuscript.* Special Collections, Shelfmark: MS. Sansk. D. 14, Bodleian Library, University of Oxford. Oxford.

Habib, I. (2008). *Medieval India: The Study of a Civilization.* India: National Book Trust.

Colebrooke, H. T. (1817, Republished 2005). *Algebra, with Arithmetic and Mensuration from the Sanscrit of Brahmagupta and Bhaskara.* New Delhi: Sharada Publishing House.

INTRODUCTION

Bagla, P. & Menon, S. (2008). *Destination Moon.* New Delhi: Harper-Collins India.

Kennedy, P. (1989). *The Rise and Fall of the Great Powers.* New York: Vintage.

Pursuit and Promotion of Science: The Indian Experience. (2001). New Delhi: Indian National Science Academy.

The Shaping of Indian Science: Indian Science Congress Association Presidential Addresses Vol. 1: 1914–1947. (2003). Hyderabad: Universities Press.

Radhakrishna, B. P. (2009). Nehru's 'Discovery of India': The Role of

Science in India's Development. *Journal of the Geological Society of India*, volume 73, number 2, pages 157–164.

Science Advisory Council to the Prime Minister of India. (2010). *India as a Global Leader in Science*, 18 September [Online]. Available at: http://www.esocialsciences.com/home/index.asp [Accessed 8 October 2010].

CHAPTER 1 BRAIN GAMES

Saini, A. (March 2010). The Rocket Principle. *GQ India* [magazine]. Many thanks to Maniza Bharucha, Managing Editor at *GQ India*, for permission to re-use material from an early draft of this article.

Krishna, A. & Haglund, E. (2008). Why do Some Countries Win More Olympic Medals? Lessons for Social Mobility and Poverty Reduction. *Economic & Political Weekly*, volume 43, number 28, 12 July, pages 143–151.

Adams, J., King, C. & Singh, V. (2009). Thomson Reuters Global Research Report: India, October 2009 [Online]. Available at: http://science.thomsonreuters.com/m/pdfs/grr-India-oct09_ago908174.pdf [Accessed 8 October 2010].

Speech by Prime Minister Manmohan Singh. (2009). Indian Science Congress. University of Kerala, Thiruvananthapuram, Kerala, India, 3 January. Available at: http://pib.nic.in/release/release.asp?relid=46369 [Accessed 1 January 2010].

Indian National Crime Records Bureau, Ministry of Home Affairs (2008) *Suicidal deaths in India* [Online]. Available at: http://ncrb.nic.in/ADSI2008/suicides-08.pdf [Accessed 3 November 2010].

Chengappa, R. (2005). Killer Exams: How to Revamp the System. *Health Administrator*, July, volume 17, number 1, pages 72–74.

National Association of Software and Services Companies. (2009). *The IT-BPO Sector in India: Strategic Review 2009*. New Delhi: NASSCOM.

Deutsche Bank Research. (2010). *The Middle Class in India: Issues and Opportunities*. Frankfurt: Deutsche Bank Research.

Prathap, G. (2005). Where Have our Young Ones Gone? The

Coolieization of India. *Current Science*, volume 89, number 7, pages 1063–1064.

Tata Consultancy Services. (2010) *Annual Report, 2009/10.* Mumbai.

CHAPTER 2 ELECTRONICS CITY

Central Intelligence Agency. *Infant Mortality Rates* [Online]. Available at: https://www.cia.gov/library/publications/the-world-factbook/ [Accessed 27 May 2010].

Upadhya, C. (2007). Employment, Exclusion and 'Merit' in the Indian IT Industry, *Economic & Political Weekly*, 19 May, pages 1863–1868.

Wooldridge, A. (2010). Special Report on Innovation in Emerging Markets, *The Economist*, 17 April.

Wharton & Boston Consulting Group. (2007). Report: 'What's Next for India: Beyond the Back Office'. Pennsylvania: Knowledge@Wharton.

Cusumano, M. & Kemerer, C. (1990). A Quantitative Analysis of US and Japanese Practice and Performance in Software Development. *Management Science*, volume 36, number 11, November, pages 1384–1406.

European Commission. (2009). EU Industrial R&D Investment Scoreboard. Brussels: European Commission. Available at: http://iri.jrc.ec.europa.eu/research/docs/2009/JRC54920.pdf [Accessed 1 April 2010].

Kulkarni, V. & Paul, S. (2009). IT firms spend more on R&D to offer new services. *The Hindu Business Line*, 20 July [Online]. Available at: http://www.thehindubusinessline.com/2009/07/20/stories/2009072050390200.htm [Accessed 1 May 2010].

Wadhwa, V., Saxenian, A., Freeman, R., Gereffi, G. & Salkever, A. (2009). *America's Loss is the World's Gain: America's New Immigrant Entrepreneurs, Part* IV. Kauffman Foundation, March [Online]. Available at: http://www.kauffman.org/research-and-policy/americas-loss-is-the-worlds-gain-americas-new-immigrant-entrepreneurs.aspx [Accessed 1 February 2010].

Saxenian, A. (2005). From Brain Drain to Brain Circulation: Transnational Communities and Regional Upgrading in India and

China, *Studies in Comparative International Development*, volume 40, number 2, June, pages 35–61.

Internet and Mobile Association of India. (2010). *I-Cube 2009 to 2010: Internet in India*, Mumbai: IAMAI. April [Online]. Available at: http://www.iamai.in/Upload/Research/ icube_new_curve_lowres_39.pdf [Accessed 5 May 2010].

Patel, N., Chittamuru, D., Jain, A., Dave, P. & Parikh, T.S. (2010). Avaaj Otalo – A Field Study of an Interactive Voice Forum for Small Farmers in Rural India, in *Proceedings of ACM Conference on Human Factors in Computing Systems, April 2010, Atlanta, Georgia, United States.* Available at: http://www.stanford.edu/~neilp/pubs/ chi2010_patel.pdf [Accessed 5 May 2010].

Arons, B. (1991). Hyperspeech: Navigating in Speech-Only Hypermedia, in *Proceedings of the third annual ACM conference on Hypertext, San Antonio, Texas, United States, 15–18 December 1991.* New York: Association of Computing Machinery. Pages 133–146.

Agarwal, S., Chakraborty, D., Kumar, A., Nanavati, A. Rajput, N. (2007). HSTP: Hyperspeech Transfer Protocol, in *Proceedings of the eighteenth conference on Hypertext and Hypermedia, Manchester, United Kingdom, 10–12 September 2007.* New York: Association of Computing Machinery. Pages 67–76.

Agarwal, S., Kumar, A., Nanavati, A. & Rajput, N. (2007). The World Wide Telecom Web Browser, in *Proceedings of the 17th International Conference on World Wide Web, Beijing, China, 21–25 April 2008*, New York: Association of Computing Machinery. Pages 1121–1128.

United States Department of Agriculture Foreign Agricultural Service. (2009). *Indian Agricultural Economy and Policy Report.* Available at: http://www.fas.usda.gov/country/India/Indian% 20Agricultural%20Economy%20and%20Policy%20Paper.pdf [Accessed 30 July 2010].

CHAPTER 3 THE LONG-LIFE BANANA

The Hindu. (2010). *Over 150 suicides in Vidarbha this year: BJP.* 19 April [Online]. Available at: http://www.thehindu.com/news/ states/other-states/article405535.ece [Accessed 20 April 2010].

REFERENCES

Dyson, T. & Maharatna, A. (1991). On the demographic consequences of the Bihar famine of 1966–67 and the Maharashtra drought of 1970–73, in *Conference on Famine and Disease, Cambridge University, July 1991*. Available at: http://repository.forcedmigration.org/pdf/?pid=fmo:741 [Accessed 1 May 2010].

Gandhi, V. P. & Namboodiri, N. V. (2006). *The Adoption and Economics of Bt Cotton in India: Preliminary Results from a Study*. Ahmedabad: Indian Institute of Management. Available at: http://www.iimahd.ernet.in/publications/data/2006-09-04_vgandhi.pdf [Accessed 1 May 2010].

Radhakrishna, B. P. (2009). Nehru's 'Discovery of India': The Role of Science in India's Development. *Journal of the Geological Society of India*, volume 73, number 2, pages 157–164.

Sadashivappa, R. & Qaim, M. (2009). Bt Cotton in India: Development of Benefits and the Role of Government Seed Price Intervention. *AgBioForum*, volume 12, number 2, pages 172–183.

Mitta, M. (2006). 3 states fix price of Monsanto's Bt cotton seeds. *The Times of India*, 1 June [Online]. Available at: http://timesofindia.indiatimes.com/india/3-states-fix-price-of-Monsantos-Bt-cotton-seeds/articleshow/1606491.cms [Accessed 1 May 2010].

Gruère, G., Mehta-Bhatt, P. & Sengupta, D. (2008). *Bt Cotton and Farmer Suicides in India: Reviewing the Evidence*. Washington DC: International Food Policy Research Institute [Online]. Available at: http://www.ifpri.org/publication/bt-cotton-and-farmer-suicides-india [Accessed 1 May 2010].

Jayaraman, K. S. (2003). US food aid to India still under GM cloud. *Nature Biotechnology*, volume 21, number 4, April, pages 346–347.

Bhole, L. M. (2008). Hind Swaraj: About the Book and Development Model, in *The Proceedings of the Seminar on Development Through Planning, Market, or Decentralization at Indian Institute of Technology Bombay, 21 January*. Mumbai: Indian Institute of Technology Bombay. Pages 87–95.

Hanstad, T., Haque, T. & Nielsen, R. (2008). Improving Land Access for India's Rural Poor. *Economic & Political Weekly*, 8 March, pages 49–56.

Stockholm Environmental Institute and Stockholm International Water Management Institute, Chalmers University. (2008). *Saving Water:*

267

From Field to Fork. Curbing losses and wastage in the food chain, 21 August [Online]. Available at: http://www.siwi.org/ documents/Resources/Papers/Paper_13_Field_to_Fork.pdf [Accessed 10 February 2010].

Damodaran, H. (2010). *Monsanto to earn Rupees 340 crore tech fee for Bt cotton.* The Hindu Business Line, 19 January [Online]. Available at: http://www.thehindubusinessline.com/2010/ 01/19/ stories/2010011951431600.htm [Accessed 21 January 2010].

McKinsey Global Institute (2010). *India's Urban Awakening: Building inclusive cities, sustaining economic growth,* April 2010 [Online]. Available at: http://www.mckinsey.com/mgi/reports/ freepass_pdfs/india_urbanization/MGI_india_urbanization_full report.pdf [Accessed 1 May 2010].

CHAPTER 4 CHARIOTS OF THE GODS

Nanda, M. (2008). Rush hour of the gods. *New Humanist*, volume 123, issue 2, March/April 2008, pages 16–19.

Bagla, P. & Menon, S. (2008). *Destination Moon.* New Delhi: Harper-Collins India.

Nanda, M. (2009). *The God Market.* Noida, Uttar Pradesh, India: Random House India.

Ravindran, K. et al. Booklet: *Indian Contributions to Science.* Kochi, India: Swadeshi Science Movement.

Mukunda, H. S., Deshpande, S. M., Nagendra, H. R., Prabhu, A. & Govindaraju, S. P. (1974). A Critical Study of the Vymanika Shastra, *Indian Institute of Science, Bengaluru* [Online]. Available at: http://cgpl.iisc.ernet.in/site/Portals/0/Publications/ RefereedJournal/ACriticalStudyOfTheWorkVaimanikaShastra.pdf [Accessed 1 May 2010].

Chakravarthy, K. (2010). No astrology here please: Bangalore University to govt. *Deccan Herald*, 12 January, page 1.

DNA India (2010) *Rationalists exorcise superstition with pongal,* 16 January [Online]. Available at: http://www.dnaindia.com/ bangalore/report_rationalists-exorcise-superstition-with-pongal_1335374 [Accessed 17 January 2010].

REFERENCES

Talk by Professor M. R. N. Murthy. (2009) *Science and pseudoscience.* National College, Jayanagar, Bengaluru, Karnataka, India, 18 November.

CHAPTER 5 THE MINDREADING MACHINE

Saini, A. (2009). Guilty. *Wired UK*, (June) [magazine]. Many thanks to David Baker, managing editor at *Wired UK*, for permission to re use material from an early draft of this article.

Alder, K. (2002). A Social History of Untruth: Lie Detection and Trust in Twentieth-Century America. *Representations, University of California Press*, number 80, autumn 2002, pages 1–33.

10th United Nations Survey of Crime Trends and Operations of Criminal Justice Systems. (2008) [Online]. Available at: http://www.unodc.org/documents/data-and-analysis/India.pdf [Accessed 1 April 2010].

Jiruska, P. et al. (2008). Clinical impact of a high-frequency seizure onset zone in a case of bitemporal epilepsy. *Epileptic Disorders*, volume 10, number 3, September, pages 231–238.

Ramachandran, V. S. & Blakeslee, S. (1999). *Phantoms in the Brain: Probing the Mysteries of the Human Mind.* London: Harper Perennial.

Morgane, P. J. (1961). Distinct 'Feeding' and 'Hunger Motivating' Systems in the Lateral Hypothalamus of the Rat. *Science*, volume 133, number 3456, 24 March, pages 887–888.

Hansen, M. (2009). True Lies. *American Bar Association Journal*, 1 October [Online]. Available at: http://www.abajournal.com/magazine/article/true_lies/ [Accessed 1 July 2010].

Kulkarni, V. & Paul, S. (2009) IT firms spend more on R&D to offer new services. *The Hindu Business Line*, 20 July.

Natu, N. (2008). This brain test maps the truth. *Times of India*, 21 July.

Raghava, M. (2008). Stop using brain mapping for investigation and as evidence. *The Hindu*, 6 September.

269

CHAPTER 6 GEEKS RULE

Barron, E. J., Harrison, C. G. A., Sloan II, J. L. & Hay, W. W. (1981). Paleogeography, 180 million years ago to the present. *Eclogae Geologicae Helvetiae*, volume 74, number 2, pages 443–470.

Biju, S. D. & Bossuyt, F. (2009). Systematics and phylogeny of Philautus Gistel, 1848 (Anura, Rhacophoridae) in the Western Ghats of India, with descriptions of 12 new species. *Zoological Journal of the Linnean Society*, volume 155, number 2, February, pages 374–444.

Yong, J. S. L. (2003) *E-Government in Asia: Enabling Public Service Innovation in the 21st Century.* Singapore: Times Media Private Limited.

Transparency International India. (2008). *TII-CMS India Corruption Study 2007: With Focus on BPL Households*, June [Online]. Available at: http://www.cmsindia.org/highlights.pdf [Accessed 1 April 2010].

Charlie, A. (2010). IT, e-governance spend may touch $4 bn next fiscal. *The Hindu Business Line*, 10 February.

Gross, G. (2009). Cloud computing, security to drive US government IT spending. *Computer World*, 9 July [Online]. Available at: http://www.computerworld.com/s/article/9135360/Cloud_computing_security_to_drive_U.S._gov_t_IT_spending [Accessed 1 July 2010].

Government of India, Department of Information Technology, Ministry of Communications and Information Technology. (2008). *Impact Assessment of e-Governance Projects*, 31 October [Online]. Available at: http://www.mit.gov.in/sites/upload_files/ dit/files/Impact AssessmentReportDraft.pdf [Accessed 1 August 2010].

Bhardwaj, M., Ohri, R. & Bajwa, H. (2009). The Ruchika Verdict. *The Indian Express*, 25 December, page 3.

NDTV correspondent (2009). India has world's largest backlog of court cases: PM. *NDTV Online*, 16 August [Online]. Available at: http://www.ndtv.com/news/india/india_has_worlds_largest_backlog_of_court_cases_pm.php [Accessed 1 August 2010].

Dharker, A. (2010). Wipe off his smirk, and that of others. *The Times of India Mumbai Edition*, 17 January.

Savage, M. (2010). Labour's computer blunders cost £26bn. *Independent* (London), 19 January.

CHAPTER 7 THE IMPOSSIBLE DRUG

NHS figures supplied by email by the British Association of Physicians of Indian Origin, 10 August 2010.

Editorial. (2009). Orphan giant. *Nature*, volume 459, 25 June, page 1034.

World Health Organization. (2009). *Factsheet: Tuberculosis Facts: 2009 Update* [Online]. Available at: http://www.who.int/entity/tb/publications/2009/factsheet_tb_2009update_dec09.pdf [Accessed 31 August 2010].

World Health Organization. (2010). *Tuberculosis: MDR-TB and XDR-TB 2010 Report* [Online]. Available at: http://www.who.int/entity/tb/features_archive/world_tb_day_2010/mdrfactsheet15mar10_19h00.pdf [Accessed 31 August 2010].

World Health Organization. (2010). *Global Tuberculosis Control: A Short Update to the 2009 Report* [Online]. Available at: http://whqlibdoc.who.int/publications/2009/9789241598866_eng.pdf [Accessed 31 August 2010].

Meli, R. et al. (2008) FishMap: A Community Resource for Zebrafish Genomics. *Zebrafish*, volume 5, number 2, pages 125–130.

Lamason, R. L. et al. (2005). SLC24A5, a putative cation exchanger, affects pigmentation in zebrafish and humans. *Science*, volume 310, number 5755, pages 1782–1786.

Gagneux, S. et al. (2006). Variable host–pathogen compatibility in Mycobacterium tuberculosis. *PNAS*, volume 103, number 8, 21 February, pages 2869–2873.

Collins, F. S. & McKusick, V. A. (2001). Implications of the Human Genome Project for Medical Science. *Journal of the American Medical Association*, volume 285, number 5, 7 February, pages 540–544.

Kitano, H. (2002). Systems Biology: A Brief Overview. *Science*, volume 295, number 5560, 1 March, pages 1662–1664.

Jayaraman, K. S. (2010). India's tuberculosis genome project under fire. *Nature.com*, 9 June [Online]. Available at:

http://www.nature.com/news/2010/100609/full/news.2010.285.
html [Accessed 30 June 2010].

BBC. (2005). India wins landmark patent battle. BBC News website,
9 March [Online]. Available at: http://news.bbc.co.uk/1/hi/sci/
tech/4333627.stm [Accessed 31 August 2010].

Jayaraman, K. S. (2009). India protects traditional medicines from
piracy. *Nature.com*, 18 February [Online]. Available at:
http://www.nature.com/news/2009/090218/full/news.2009.107.
html [Accessed 31 August 2010].

Gupta, A. (2010). Scientists want changes in innovation Bill.
Livemint.com, 8 February [Online]. Available at:
http://www.livemint.com/2010/02/07225403/Scientists-want-
changes-in-inn.html [Accessed 31 August 2010].

CHAPTER 8 BRAINPOWER

World Nuclear Association. *Nuclear Century Outlook Data* [Online].
Available at: http://www.world-nuclear.org/
outlook/nuclear_century_outlook.html [Accessed 17 August 2010].

World Nuclear Association. *Data on Thorium* [Online]. Available at:
http://www.world-nuclear.org/info/inf62.html#LWBR [Accessed
17 August 2010].

World Nuclear Association. *Nuclear Power in India* [Online]. Available
at: http://www.world-nuclear.org/info/inf53.html [Accessed 17
August 2010].

Jayaraman, K.S. (2010). India's nuclear future. *Nature.com*, 4 January
[Online]. Available at: http://www.nature.com/news/
2010/100104/full/news.2010.0.html [Accessed 1 September 2010].

Grimes, R. W. & Nuttall, W. J. (2010). Generating the Option of a Two-
Stage Nuclear Renaissance. *Science*, volume 329, number 5993, 13
August, pages 799–803.

Martin, R. (2009). Uranium Is So Last Century – Enter Thorium, the
New Green Nuke. *Wired.com*, 21 December 2009 [Online].
Available at: http://www.wired.com/magazine/
2009/12/ff_new_nukes/ [Accessed 1 September 2010].

IAEA. (2005). Report: *Thorium fuel cycle — Potential benefits and
challenges*. Vienna: IAEA [Online]. Available at http://

www-pub.iaea.org/mtcd/publications/pdf/te_1450_web.pdf
[Accessed 1 September 2010].

Indian delegation. (2008). Extending the global reach of nuclear energy through thorium presented at the *IAEA 52nd General Conference of Member States*, Austria Center, Vienna, 29 September –4 October.

Jagannathan, V. (1999). A Thorium Breeder Reactor Concept for Early Induction of Thorium with No Feed Enrichment. *Bhabha Atomic Research Centre newsletter*, number 187, August, pages 176–182.

Sinha, R. & Kakodkar, A. (2010). India's passive breeder. *Nuclear Engineering International*, 17 May [Online]. Available at: http://www.neimagazine.com/story.asp?storyCode=2056393 [Accessed 1 August 2010].

CHAPTER 9 ROCKET SCIENCE

Times of India. (2010). *ISRO's budget is just three per cent of that of NASA*, 3 March.

de Selding, P. B. (2010). Number of Worldwide Space Agencies on the Rise. *Space.com*, 25 February [Online]. Available at: http://www.space.com/news/worldwide-space-agencies-on-the-rise-sn-100224.html [Accessed 10 September 2010].

SCImago (2007) Science publication citation rankings by country. *SCImago Journal & Country Rank* [Online]. Available at: http://www.scimagojr.com [Accessed 7 October 2010].

Evidence, a division of Thomson Reuters. (2010). *Bibliometric study of India's research output and international collaboration: A report commissioned on behalf of Research Councils UK*. June [Online]. Available at: http://www.india.rcuk.ac.uk/cmsweb/downloads/rcuk/india/BibliometricstudyIndiaresearchoutput.pdf [Accessed 1 August 2010].

ACKNOWLEDGEMENTS

Travelling through India alone isn't easy, so I have more people to thank than I can fit in a page. But I couldn't have started this journey without my agent, Peter Tallack, or Rupert Lancaster, Kate Miles and the team at Hodder, or Thomas Abraham (another geek) and his team at Hachette India.

For logistical support, thanks to Raj Kumar Sharma at the Bhabha Atomic Research Centre, Vinod Scaria at the Institute of Genomics and Integrative Biology, Seetha Srikanth at Tata Consultancy Services, Jaya Nair at the Indian Science Congress, and makemytrip.com. At the British and American ends, I was kindly helped by James Corbett, Brian Clegg, David Castle and Patrick Ion at the American Mathematical Society, Gillian Evison at the University of Oxford, Dave Sellwood at University College London, Amanda Brown at Queen Mary University of London, Robin Grimes at Imperial College London and Kim Plofker at Union College. For their generous time proofreading, thanks to Samir Sheldenkar, Ruta Nimkar, Samantha Haque and Rima Saini.

Thanks above all to my friends and family in New Delhi, London, New York and Neath, who became fans of *Geek Nation* before it was even finished.

Finally, Mukul, let's park the boat now and go for a burger. It's on me.

INDEX

275

SITE (Satellite Instructional Television Experiment) 15
Sivasubbu, Sridhar 209–10, 211–12, 214, 219, 225
solar energy 119, 261
South Korea 175
Soviet Union 12
space programme 3–16, 47, 254–9
space travel, ancient 120–1
Spoken Web 65–72
student stress 37–8
Sun Microsystems 64
superstition 12, 117, 125, 126, 129, 139–40, 142, 143
Swaminathan, M. S. 107
systems biology 226–8

Taiwan 175
Tata Consultancy Services 40–1, 42–3, 44–6, 48, 50, 53, 62, 175, 183
Technocracy 76–7
terrorism 147
Tesla, Nikola 220
Thackeray, Vinod 110–11
Thiruvananthapuram 234, 250, 259
Thomas, Zakir 220–3, 224, 227, 229, 231–2
thorium/thorium reactors 240–2, 243, 244, 245–7, 246, 249
Thumba 3–5
Tiwari, Radhe Shyam 21
tomatoes, transgenic 105–6
TringMe 73–5
truth machines 145–7, 149, 150–7, 159–64, 167

tuberculosis 200–9, 215–17, 227, 228–9, 231–2
Tuberculosis Research Centre 203–7, 214–16
turmeric 117

United States of America
engineering 12; Indians working and training in 63–4; R&D budget 18
Upadhya, Carol 59
Upadhyaya, Shripati 161
uranium 235–6, 238, 239, 241, 245, 246, 248
USB (Universal Serial Bus) 64

Vaimanika Shastra 114–16, 119, 135–6
Vedas 114, 118–19, 121, 135, 136, 140–1, 142–3
Vedic science 114–16, 117, 118–44
Vidarbha 85–9, 91–2, 110–12
Vikram Sarabhai Space Centre 3–5, 254
Voice PHP 74

Wadhwa, Vivek 64
Watson, James 83
Western Ghats 169–74, 179 *see also* Lavasa
Wipro 175, 176, 179, 183
Wrighton, Scot 172, 174, 177

Y2K bug 41–2, 43
Yahoo 65

zebrafish 209–11, 212